Apex Anchor: Full-Service Logistics Transportation Node at Top of Gravity Well

Study Lead:
Paul W. Phister, Jr,, Ph.D., P.E.

Editors:
Paul Phister, Ph.D., P.E.
Pete Swan, Ph.D.
Sandra Therrien
Joyce Efferdink

Authors:
Paul W. Phister, Jr., Ph.D., P.E.
Joyce Elferdink
John Knapman, Ph.D.
Bassem Sabra, Ph.D.
Pete Swan, Ph.D.
Sandra Therrien

Prepared for the
International Space Elevator Consortium

August 2024

International Space Elevator Consortium   *ISEC Position Paper #2024*

Apex Anchor: Full-Service Logistics Transportation Node at Top of Gravity Well

Copyright @ 2024 by
Paul W. Phister, Jr.
International Space Elevator Consortium

All rights reserved, including the rights to reproduce
This manuscript or portions thereof in any form.

Published by Lulu.com

info@isec.org

Cover Image by Swan: Portions of the Climber image -- <a href="https://www.freepik.com/free- vector/sticker-set-outer-space-objects astronauts_28458273.htm#page=2&query=spacecraft%20drawing&position=39&from_view=keyword&track=ais">Image by brgfx</a> on Freepik

978-1-304-17909-8

Printed in the United States of America

International Space Elevator Consortium   *ISEC Position Paper #2024*

Preface

The old asteroid on the end of a rope idea has gone the way of carriages drawn by horses. Once the Modern-Day Space Elevator emerged in 2020, the concept of an active transportation node at the Apex Anchor became obvious. The enhancement of the roles at an Apex Anchor showed how it will become a Full-Service Transportation Node for several significant missions centered around the "Logistics Center at the top of the Gravity Well." The capability to raise all manner of space system components and segments 100,000 km using electricity – as a Green Road to Space – enables assembly of major space missions only dreamt of before. The inherent rapid release velocity (7.76 km/sec) at an Apex Anchor – and the ability to release every day of the year towards any destination, such as the Moon, Mars and even Pluto – opens up the universe to humanity. These capabilities at the Full-Service Transportation Node led to enabling of missions such as:

- A release point for interplanetary science missions with space systems of any size (assembled at the top of the gravity well) to any planet with daily windows.
- Interplanetary human missions (after development of full Operational Capability) with space systems of any size to any planet with daily windows.
- Interplanetary full-service logistics, storage and resupply capability with "just in time delivery"
- Cis-Lunar full-service logistics, storage and release for supply missions with "just in time delivery"
- Astronaut Rescue staging area with storage of necessary rescue material such as oxygen, habitats, water, rocket fuel, power, and food. (only 14 hours from the Moon)
- Planetary Defense asteroid detection sensors, with multiple 100,000 km tethers on each side of the Earth (206,000 km baseline) for stereoscopic vision of incoming asteroids from the Sun or out of the asteroid belts. In addition, the storage of planetary defense spacecraft would enable rapid response to near-term threats within 24 hours. Flexibility of responses can be accomplished by storing, then rapid assembly of various defensive segments of planetary protection space systems, depending upon the threat.

These capabilities also lead to a vision that is revolutionary and illustrates the impacts of these transformational characteristics. The Apex Anchor is a Full-Service Transportation Node driving Space Elevators' vision:

> *Space Elevators are the Green Road to Space while they enable humanity's most important missions by moving massive tonnage to GEO and beyond. This is accomplished safely, routinely, inexpensively, daily, efficiently, and they are environmentally neutral.*

Acknowledgements

Having a vision for going outworld is the next step leading to programs and capabilities that realize dreams. The future of exploring space has been accelerating over the past 50 years. However, the last 10 years have shown how hard it is to accomplish space-based missions. This study considered the future of the Apex Anchor and possible missions that can be accomplished when humans begin to inhabit the Apex Anchor. The International Space Elevator Consortium (ISEC) brought in a special team of international experts to investigate potential missions that can be accomplished aboard the Apex Anchor.

The Team:

Study Lead and Author: Paul W. Phister, Jr., Ph.D., P.E., is President of MANIAC Consulting and holds a professional engineering license in both software and electrical engineering from the State of Texas. He is currently an Adjunct Professor at the Mohak Valley Community College (MVCC) and Herkimer Community College. Additionally, Dr. Phister spent 25 years in the military (Lt Col., retired) where he worked primarily in space systems development, acquisition, and operations.

Author and Editor: Joyce Elferdink has written one self-published book, and considers *Battle of Jericho 2038,* that uses a space elevator as a backdrop, to be a sequel. She has a master's degree in communication and urban studies and teaches communications and English Composition courses at a local university. She has also been a banking branch manager, an economic development director and a Peace Corps volunteer in Kazakhstan and Ukraine. Her love of science fiction began when assigned to read C.S. Lewis' Space Trilogy. His writing taught her that creativity is necessary to look into the future and suggest opportunities for a better world. That has become her personal goal and expectation for the space elevator program.

Author: John Knapman, Ph.D., is Director of Research for the International Space Elevator Consortium. He holds a degree in mathematics and a doctorate in Artificial Intelligence. His work on the space elevator has included analysis of the conditions in the atmosphere and methods for space elevators to deal with them, ways to reduce the strength requirement of the tether material, and opportunities to exploit and augment the high speed at the apex anchor to launch interplanetary spacecraft. He has several publications in Acta Astronautica and JBIS.

Author: Bassem Sabra, Ph.D., Professor of Physics & Astronomy and Department Chairperson at Notre Dame University – Louaize (NDU), Lebanon. His field of research is observational astrophysics (supermassive black holes and active galactic nuclei: using multi-wavelength spectroscopy, modeling, and machine learning to study accretion physics, feedback, and the co-evolution with the host galaxy).

Author and Editor: <u>Peter A. Swan</u>, *Ph.D.*,
He is Past President of the International Space Elevator Consortium, ISEC's Chief Architect, and Board Member of Space Elevator Development Corporation. As such, he leads teams who further the Space Elevator concept with incremental studies and yearly conferences. Over the last 20 years he has published many books on the topic as author, co-author, and/or co-editor. He is a full member of the International Academy of Astronautics (IAA). Pete received his Ph.D. from the University of California at Los Angeles in Mechanical Engineering. He is a Fellow of the BIS and the AIAA.

Author and Editor: <u>Sandra Therrien</u> is an aerospace professional who has been in the business for 30 years in various roles from communications and navigation systems journeyman and flightline mechanic in the U.S. Air Force, to technical author of maintenance manuals for unmanned and experimental aircraft, to working in the mission control center for flight test of SpaceShipTwo Unity, to program and project management for aircraft development and flight test programs. She is a backcountry pilot, having taken bush pilot training in Talkeetna, Alaska, and authors speculative science fiction novels for fun.

As in all projects, it takes a team. We wish to thank them for their contributions and their belief regarding the future of the Apex Anchor. This is a necessary step in the evolution of the Space Elevator system.

International Space Elevator Consortium ISEC Position Paper #2024

Executive Summary

Space Elevators position humanity to address Earth's challenges from a new vantage point. We are on the brink of transforming our relationship with space, offering an eco-friendly, cost-effective, and efficient logistics method to transport large cargoes into space. This space port will provide unparalleled opportunities in space exploration, resource utilization, and satellite assembly. Starting in the late 2030s, Space Elevator infrastructures will deliver satellites and other payloads to GEO, the Moon and Mars at the rate of 30,000 tonnes, every year. Once the Modern-Day Space Elevator development program emerged in 2020, the concept of an active transportation node at the Apex Anchor became obvious. The enhancement of the roles at an Apex Anchor showed how it will become a Full-Service Transportation Node focused around the "Logistics Center at the top of the Gravity Well." The inherent rapid release velocity (7.76 km/sec) at an Apex Anchor – and the ability to release every day of the year towards any destination, such as the Moon, Mars and even Pluto – opens up the universe to humanity. These capabilities at the Full-Service Transportation Node led to the enabling of missions such as:

a. **Apex Anchor Operations Center:** Will support all customers in day-to-day operations

b. **Planetary Defense:** Developed to be a rapid response capability to "Protect the Planet"

c. **Space Transportation Port (a.k.a. "Truck Stop"):** Will provide facilities to accept space systems entering the Apex Anchor [from orbits or from the tether], service the vehicles, and release back to orbit or along the tether

d. **Space Construction Platform:** Will support and enable construction of facilities and space systems with ability to build, repair, improve upon the Apex Anchor

e. **Space Hospital and Rehabilitation Center:** Attend to the medical needs of the residents and the transient personnel upon need.

f. **Space Logistics, Storage and Distributing Center:** Enable the operations of multiple missions upon the Apex Anchor with storage and distribution of logistics.

g. **Space Solar Power Distribution Center:** Enable the collection of solar energy and then the ability to distribute around the Apex Anchor and to regional space systems in need of electrical energy.

h. **Space Astronomical Observatory:** Enable quiescent location for observation of the arena around the Apex Anchor and the solar system and universe as needed.

i. **Nuclear Waste Disposal System:** Allow nuclear waste from the Earth to transit the Apex Anchor on its way from the tether to a destination (orbit) far away from the Earth for terminal storage of high-level nuclear waste

j. **Space Hotel:** Enable transients and residents to have living quarters within the Apex Anchor that are safe and convenient for the work environment.

k. **Next Generation International Space Station:** Empower multiple countries to participate in a space station at the Apex Anchor that would enable a stable environment for government & commercial operations and existence.

Doors to many jobs for adventurous people will be open to fulfill these roles in space. In the *Space Elevator Near Term Vision* section readers were asked to imagine being on a trip into space. Now we ask you to go beyond that image and think of what part you could play after the Space Train has brought thousands of travelers to the Apex Anchor, some to stay. Imagine how you could share your expertise and earn a good living either in commercial jobs or in some official capacity, maybe while living in a Martian colony. There will surely be a place for engineers, hospitality and health care workers, construction personnel, even astronomers, climate entrepreneurs and business and IT specialists. And of course, scientists, researchers' teachers and writers (to publicize ways to join the growing numbers who want to make a difference and believe they can do that most effectively in space, either remotely or physically).

Table of Contents

Preface iii
Acknowledgements iv
The Team: iv
Executive Summary vi
Table of Contents viii
Chapter 1:   Introduction and Status of Space 1
    1.0   Imagination 1
    1.1   Space Elevator Near Term Vision 1
    1.2   Overview - "The Future Starts Today, Not Tomorrow" 2
    1.3   Elevate 3
    1.4   What is a Space Elevator? [ISEC Website, 2024] 4
    1.5   Integrated Approach 6
    1.6   What is a Galactic Harbour? [ISEC Website, 2024] 7
    1.7   Major Operational Characteristics 8
    1.8   Growth of Space Elevators 9
    1.9   Baseline Infrastructures: 9
    1.10 ISEC Apex Anchor Study Starting Point 10
    1.11 Major Questions for the Study 10
    1.12 Chapter Breakout 11
Chapter 2:   Apex Anchor as a Full-Service Transportation Node 12
    2.0   Introduction 12
    2.1   Modern-Day Space Elevator 13
    2.2   Permanent Space Access Infrastructure 14
    2.3   ISEC Study Starting Point 14
Chapter 3:   Velocity Enhancements 19
    3.1   Introduction 19
    3.2   Critique 22
    3.3   The Celestial Sphere 23

3.4	Slingshot	23
3.5	Calculations	25
3.6	Secondary Tethers	26
3.7	Description	27

Chapter 4: Apex Anchor Missions 31

4.0	Introduction	31
4.1	Apex Anchor Operations Center	31
4.2	Planetary Defense	31
4.3	Space Transportation Port (a.k.a "Truck Stop")	33
4.4	Space Construction Center	34
4.5	Space Hospital	35
4.6	Space Logistics, Storage and Distribution Center	35
4.7	Space Solar Power Distribution Center	38
4.8	Apex Anchor Astronomical Observatory	39
4.9	Hi-Level Nuclear Waste Disposal System	42
4.10	Space Hotel at the Apex Anchor [Mafi, 2023]	45
4.11	Next Generation International Space Station	45

Chapter 5: Apex Anchor Mission Examples 47

5.0	Introduction	47
5.1	Reference Domains	47
5.2	Detailed Mission Examples	48
5.3	Space Hospital at the Apex Anchor	63
5.4	Space Hotel at the Apex Anchor	68

Chapter 6: Study Summary, Conclusions and Recommendations 72

6.1	Summary	72
6.2	Conclusions	73
6.3	Recommendations	75

References 76

Appendix A: Glossary 81

Appendix B: International Space Elevator Consortium 83

Appendix C: Modern Day Space Elevator Body of Knowledge (www.isec.org) 85

Chapter 1: Introduction and Status of Space

1.0 Imagination

It is the writer's "duty to indicate the dividing line between imagination and reality." Arthur C. Clarke wrote this in *The Fountains of Paradise* in 1978 (p. 298). [Clarke,1978] The next few visionary paragraphs look at the Space Elevator as a delightful trip up to the Apex Anchor and on to parts beyond as a starting point to fully absorb the strengths of this new concept – a Full Service "truck stop" on the way to the Moon and Mars. The authors wish to quickly illustrate the permanent infrastructure for a transportation system into outer space.

> The Modern-Day Space Elevator is made up of climbing cars perched on ribbons (also known as cables or tethers) shaped into vertical railroad tracks that could move from the Earth's surface and soar vertically to a counterweight 100,000 km up. Clarke imagined it as a bridge to the stars that could move millions of people and millions of tons of materials into the sky driven by cheap electricity making it more cost effective, plus safer and faster than rockets. Another scientist has described it as looking "something like Jack's Beanstalk, anchored in the ground with a stalk reaching through the clouds." [Naumann, 2023] A few years later Clarke wrote, "The space elevator is an idea whose time has come," a point he made referring to other authors who independently reinvented the idea at least three times in the 70s. Imagine that it is 2045 and you have just purchased a ticket to visit the moon. It's expensive but you've been waiting for the cost to shrink and now it's a 99% savings over a seat on a rocket. At the Earth port in the middle of the Pacific, you have climbed aboard one of several vehicles called climbers that look something like a capsule or can with a pencil point at each end, and with the amenities of a train car. Your initial journey delivers you to GEO orbit, a circular geosynchronous orbit 35,786 km or 22,236 miles in altitude above Earth's equator. Much is happening at GEO; shops are springing up and commercial businesses and warehouses are supplying the developments at the Apex Anchor as well as the planned settlement on the Moon. Once the climbers are stocked, you reboard. From there, you will be traveling approximately 250 miles per hour to reach the Apex Anchor (the transfer node for your trip) in ten days. Think of the view from your window as you climb 62,000 miles (100,000 km) above the Earth's surface and eventually gaze upon the blue marble. Once you reach the Apex Anchor, you will have a few days to relax and send gifts home before you transfer to the Lunar transport vehicle which will deliver you to the Lunar surface, or mission complete.

1.1 Space Elevator Near Term Vision

You may now terminate your imaginative trip into space in a space elevator because in just twenty-five years (2049) there well may be a space elevator that transports people and materials as far as the Apex Anchor (100,000 km) and in then send them on to the moon and beyond. Space elevators are no longer just a figment of an authors' imagination. They are about to become reality, and in the future will allow millions of people to ride in safety and comfort to the moon, Mars, and other planets. Here are some of the plans being implemented by the International Space Elevator Consortium (ISEC) to make it a reality:

> Fast Release Concept: The total Space Elevator Apex Anchor release concept is fast enough to allow travelling around our solar system within months instead of decades. However, there is more for those seeking answers beyond Earth – how about the ability to send multi-hundreds of tonnes along this path in a huge space system with fuel for rendezvous and landing along the way? How? Forget the Rocket Equation! You assemble a large spacecraft above the gravity well (or pull of gravity) of the Earth, add velocity to it by virtue of the Earth's rotation and then release it with extra velocity as if from a giant slingshot. An extra-long space elevator could achieve these goals by 2038.

We have been considering what may be available 25 years from now, but as Pope John Paul II stated, "The future starts today, not tomorrow."[Pope John Paul II] The future of space elevators has already begun at the top-level analysis level leading to the vision that the Apex Anchor is where the Space Elevator will meet the shoreline of outer-space and leads to where the transportation story of the 21st century will meet the final frontier." (Swan, 2023)

1.2 Overview - "The Future Starts Today, Not Tomorrow"

When looking at the future and including Space Elevators, it seems fortunate to create a Space Elevator vision and lay out the future along those lines with a top-level approach and a list of strengths. The current version is as follows:

> ***Vision:*** *Space Elevators are the Green Road to Space while they enable humanity's most important missions by moving massive tonnage to GEO and beyond. This is accomplished safely, routinely, efficiently, inexpensively, daily, and they are environmentally neutral.*

Assembly at the Top of the Gravity Well: This ISEC research report developed the concept of assembly, refuel, repair or full construction of large space systems above the gravity well. This concept is unique as each of the segments arrives at the Apex Anchor with tremendous energy gained by rising from the ocean surface. In essence, each segment is similar to a fully fueled segment of a truck convoy resulting in a Full-Service Transportation Node. The concept – let's build huge space systems that already have monumental energy stored by being out of the gravity well and having huge velocities upon release. The study continued to develop the concept and came up with a remarkable number of missions that would be revolutionary in concept, but evolutionary in off planet.

In addition to vision, reality must be addressed in parallel. Now -- when we look at the Moon and dream of spaceflight, we forget how extremely difficult it was to accomplish, both in energy and design complexity. Tsiolkovsky's rocket equation consumes so much mass to achieve orbit that, historically, we have been restricted - as to size and weight - what we can deliver. This results in the conundrum of rockets:

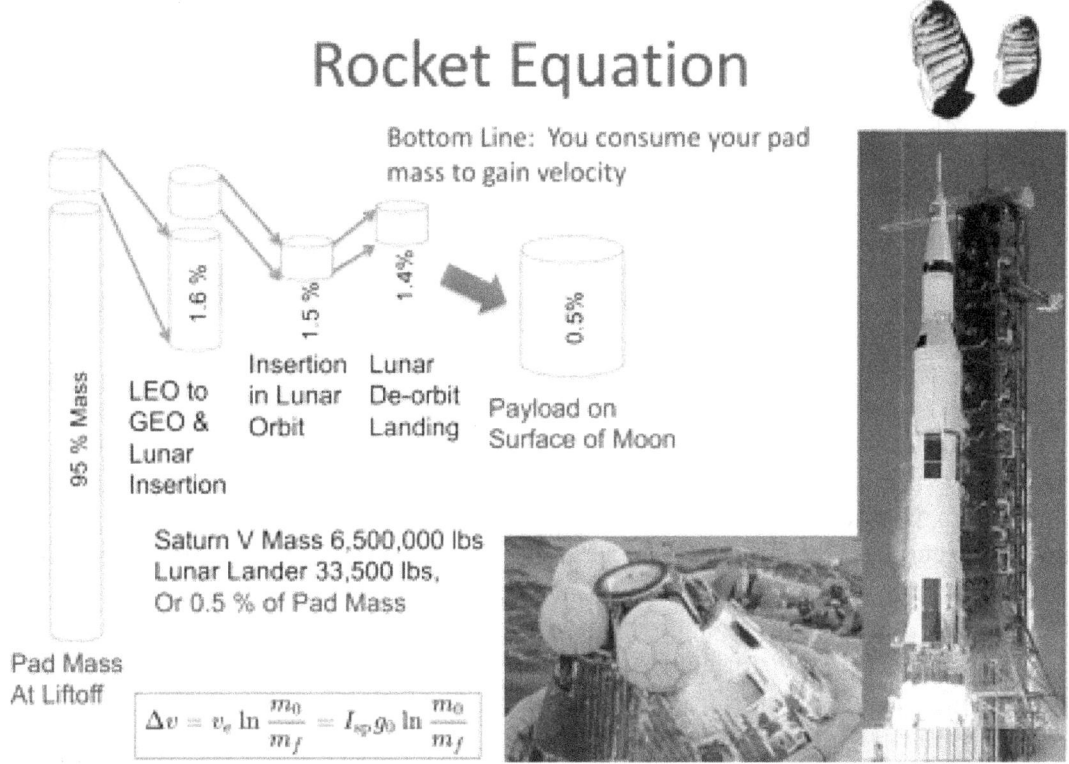

Figure 1.2.1: Rocket Conundrum, the Apollo Example (ISEC Image)

"The conundrum of rockets is the simple realization that the delivery of mass to its destination is an insignificant percentage of the mass on the launch pad. The question is why you would employ a methodology for delivery that only delivers less than one percent to your desired location (let's say the future gateway around the moon.)." [Swan, 2022] This case is illustrated in Figure 1.2.1, the Apollo example, shows that only 0.5% of the pad mass at Cape Canaveral was delivered to the surface of the Moon with the rest consumed or thrown away. Yes, reusability improves costs and reliability, but it does NOT improve efficiency of logistics delivery.

1.3 Elevate

The strengths of Space Elevators lead to the term *ELEVATE!* Which leads to the realization that the strengths of a permanent space access infrastructure provide a break-out capability for humanity's needs from and in space.

1.3.1 Top Level Approach

Space Elevators position humanity to address Earth's challenges from a new vantage point. We are on the brink of transforming our relationship with space, offering an eco-friendly, cost-effective, and efficient logistics method to transport large cargoes into space. This gateway will provide unparalleled opportunities in space exploration, resource utilization, and satellite assembly. Starting in the late 2030s, Space Elevator infrastructures will deliver satellites and other payloads to GEO, the Moon and Mars at the rate of 30,000 tonnes, every year. This surpasses the total launched between 1957 and 2022. Indeed – a seismic shift! By harnessing electricity for lift, each space elevator promises daily deliveries of up to 14 tonnes to geostationary orbit (GEO), dramatically reducing the environmental impact as compared to rocket launches. Space Elevator designs have an unmatched 70% pad mass to GEO efficiency, as compared to only 2% for rockets. They have the potential to unlock solutions to Earth's most pressing challenges such as harvesting solar power from space, climate monitoring, and global communication networks. As humanity stands on the cusp of this new era, these ribbons from ocean to space offer the promise of making space accessible to all, fostering global cooperation, positioning humanity to address Earth's challenges, and inspiring a sustainable future for our planet. [ISEC Webpage, 2024]

These projections of capabilities for Space Elevators, when they become operational in the late 2030's, will indeed support the current movement of humanity to the Moon and then on to Mars in a governmental and commercial effort driven by visions of individuals as well as the desires of countries. It seems that the future should incorporate the proposed "transportation story of the 21st century," [Fitzgerald, 2020] or the Space Elevator. This transportation infrastructure would have the strengths and characteristics of bridges and train tracks, or in other words: Full-service transportation infrastructure with space ports for assembly, repair, construction, refueling and release along the way. From a historical transportation perspective, canals, channels and deep-water ports are infrastructure while ships are vehicles. Likewise, interstate highways, bridges, and trans-continental rail systems are infrastructures for ground transportation while trucks and trains are the vehicles that use it. International airports and related facilities are infrastructures for air travel while planes are vehicles. From this perspective, rockets, no matter how large and reusable they may be in the future, will always be vehicles - not a permanent space infrastructure. When one evaluates permanent transportation infrastructures, several strengths surface that enable missions that are difficult today with rockets.

These will be explained throughout this report and missions will be illustrated that leverage these unique capabilities of Space Elevators. They include: daily, routine, unmatched logistics efficiencies [70% of pad mass to GEO and beyond], unmatched logistics mass to orbit [30,000 tonnes to GEO and beyond per year once Space Elevators reach initial operational Capability (IOC)] and environmentally neutral operations leading to the concept of a Green Road to Space.

1.4   What is a Space Elevator? [ISEC Website, 2024]

The Space Elevator is a transportation infrastructure that is ecological and "beats gravity well." Its overriding strength is that it supplies massive amounts of cargo to GEO and beyond in an environmentally friendly manner. By using electricity to raise its payloads to "toss" towards their destinations, the Space Elevator is a "Big Green Machine." It not only does not consume fossil fuels to raise itself, but it also enables tremendously difficult missions that are not fully realizable with traditional rockets (or even the newer reusable ones). The Space Elevator Transportation System consists of six major segments as shown in Figure 1.4.1.

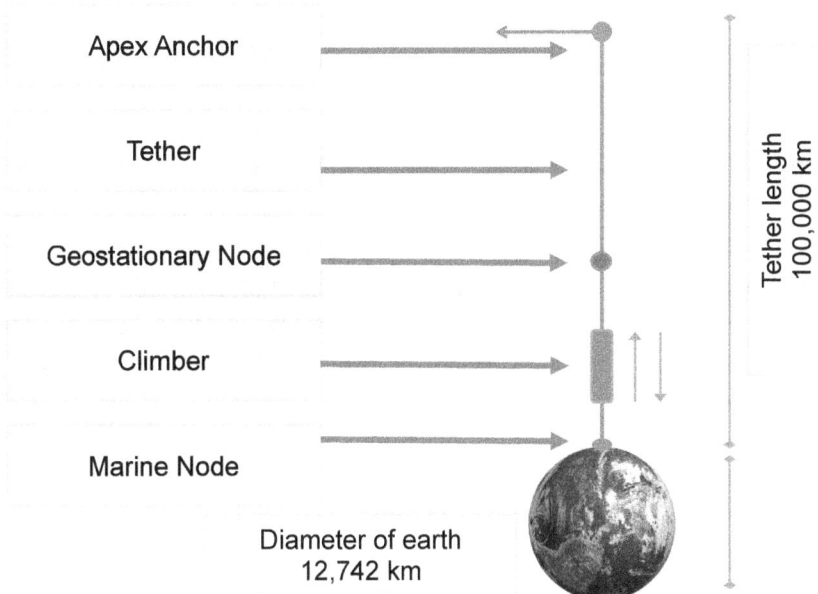

Figure 1.4.1: Space Elevator Architecture (ISEC image)

- Earth Port and Earth Port Region: A complex located at the Earth terminus of the tether, on the ocean at the equator to support its functions. These mission elements are spread out within the Earth Port Region. When there are two or more termini of tethers, the Earth Port reaches across the region and is considered one Earth Port. The volumetric region around each Earth Port includes a space elevator column for each tether and the area between multiple tethers when they operate together. The Earth Port Region will include the vertical volume through the atmosphere up to where the space elevator tether climbers start operations in the vacuum and down to the ocean floor

- GEO Region: The complex of Space Elevator activities positioned in the Space Elevator GEO Region of the Geosynchronous belt [36,000 kms altitude]; directly above the Earth Port. There will be several sub-nodes; one for each tether, one for a central main operating platform, one for each "parking lot," and others. Encompasses all volume swept out by the tether around the Geosynchronous altitude, as well as the orbits of the various support and service spacecraft "assigned" to the GEO Region. When two or more space elevators are operating together, the region includes each and the volume between elevators.

- Apex Anchor and Region: A complex of activity is located at the end of the Space Elevator providing counterweight stability for the space elevator as a large end mass. Attached at the end of the tether will be a complex of Apex Anchor elements such as reel-in/reel-out capability, thrusters to maintain stability, command and control elements, etc.. Release from Apex Anchor enables "free flights to Mars and beyond" rapidly and daily. The region is the volume swept out by the end of the tether during normal operations. When two or more space elevators are operating together, the region spreads to the volume between.

- Tether Climber: Vehicles able to climb or lower themselves on the tether, as well as releasing or capturing satellites for transportation or orbital insertion. Estimate at Initial Operational Capability is 14 tonnes of payload per day. At Full Operational Capability, that moves to 79 tonnes per day – per Space Elevator.

- Tether: 100,000 km long woven ribbon of Space Elevator with sufficient strength to weight ratio to enable the concept. Recent materials such as super laminate graphene (single crystal graphene or white Graphene) have been shown in the laboratory to be both long enough (produced over 20-meter lengths) and strong enough (greater than 150 GPa). They will be ready within 5-10 years for our 2037 capabilities. [Wright, 2023]

- Headquarters and Primary Operations Center: This facility for business and mission operations will have a main location other than at Earth Port, more likely near Space Elevator Access City.

As the development of Space Elevators progressed, it became obvious to the architects that it really is the transportation story of the 21st century. As such, a larger view of the future operations would include two Space Elevators working cooperatively with one road up and one road down – or both up to maximize cash – or a mix of the two. This reliable, routine, safe, and efficient access to space is close at hand with the Initial Operational Capability during the late 2030's. This overall concept is called the Galactic Harbour and has some very beneficial aspects as it develops into an essential part of the global and interplanetary transportation infrastructure.

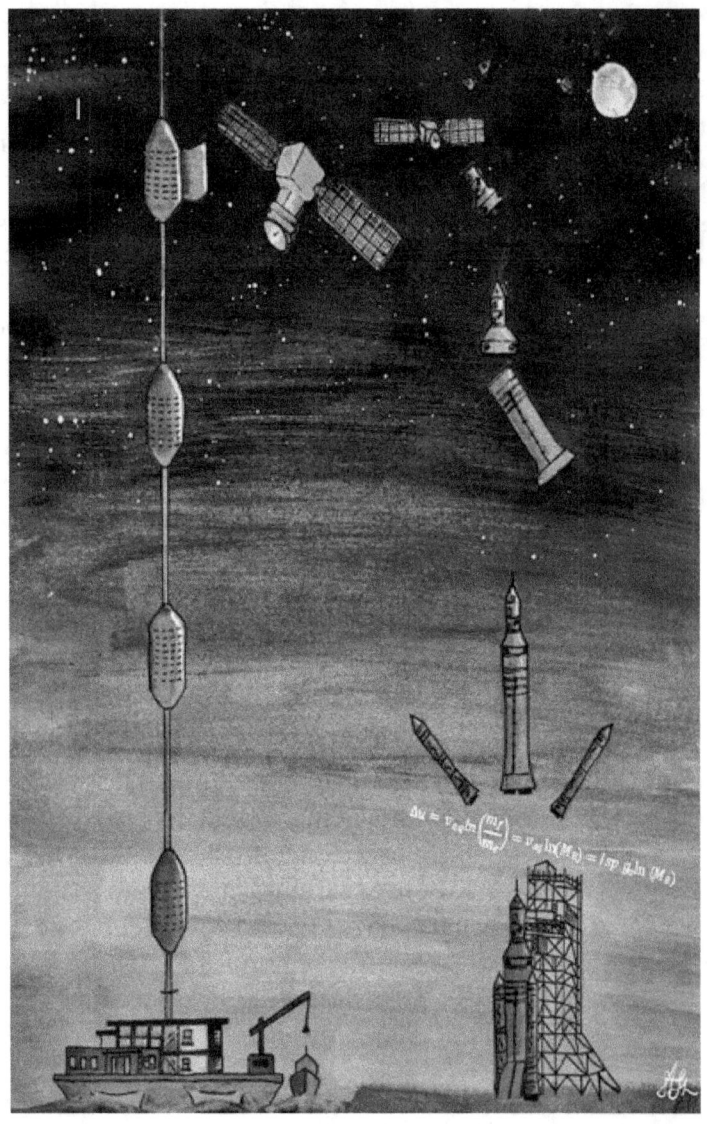

Figure 1.5.1: Space Elevators Team with Rockets (Stanton Image)

1.5 Integrated Approach

A Dual Space Access Strategy combines the strengths of two methodologies to reach space, advanced rockets and Space Elevators [see Figure 1.5.1]. This combination of strengths enables humanity to achieve great missions through cooperation and competition. This approach relies on Space Elevator's traditional strengths such as inexpensive, safe, daily, routine, unmatched efficiencies, unmatched velocities and unmatched movement of mass while having special characteristic of Earth friendly, and its ability to avoid the rocket equation. Operational strengths of rockets, with future growth, ensure rockets reach multiple orbits [specifically Low Earth Orbit] and enable rapid movement through the radiation belts.

1.6 What is a Galactic Harbour? [ISEC Website, 2024]

The Space Elevator is a transportation infrastructure that is ecological and "beats gravity well." Its overriding strength is that it supplies massive amounts of cargo to GEO and beyond in an environmentally friendly manner. By using electricity to raise its payloads to "toss" towards their destinations, the Space Elevator is a "Big Green Machine." It not only does not consume fossil fuels to raise itself, but it also enables tremendously difficult missions that are not fully realizable with traditional rockets (or even the newer reusable ones). The Space Elevator Transportation System consists of six major segments (see Figure 1.4.1): As the development of Space Elevators progressed, it became obvious to the architects that it really is the transportation story of the 21st century (see Figure 1.6.1). As such, a larger view of the future operations would include two Space Elevators working cooperatively with one road up and one road down – or both up to maximize cash – or a mix of the two. This reliable, routine, safe, and efficient access to space is close at hand with the Initial Operational Capability during the late 2030's. This overall concept is called the Galactic Harbour and has some very beneficial aspects as it develops into an essential part of the global and interplanetary transportation infrastructure.

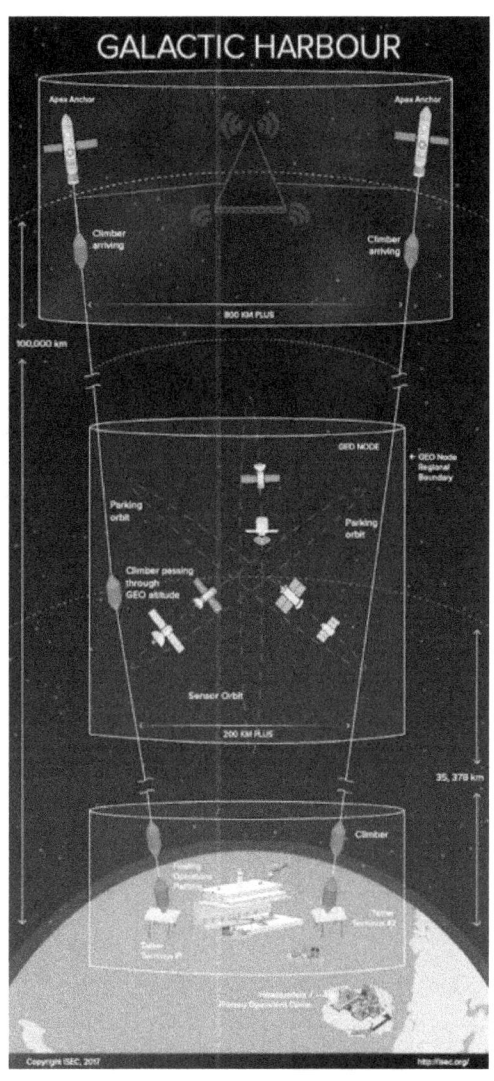

Figure 1.6.1: Galactic Harbour Layout (ISEC Image)

One capability that enables dreams of the future is the ability to release from unlimited locations along Space Elevators – which will allow high speed launches for transit throughout our solar system. This study is analyzing the capabilities and missions that are going to be orchestrated at the Apex Anchor, the highest reach of the vertical tether. When humanity does move off planet, the need for efficient logistics, with the concept of on-time delivery, will dominate logistics movements. This "train like station" at 100,000 km altitude will enable unique mission success because of its capability to fulfill the needs for missions requiring transportation and logistics support, [especially at low cost, unmatched deliver efficiencies, and routinely/daily]. One of the big puzzles of this study is to create and understand missions that cannot be achieved today because of the rocket equation. This will be discussed across two chapters in this report.

1.7 Major Operational Characteristics

There are three major approaches for the Space Elevator to leverage when they become operational filling huge goals for so many future missions. They are:

- Green Road to Space: [Eddy, 2021] When evaluating the operations of Space Elevators, the reality surfaces that they are operationally neutral in that they raise their tether climbers with electricity and do not impact the space around them. They also do not leave space debris along their path as they ascend towards the Apex Anchor. These transportation infrastructures become the green road to space!

- Dual Space Access Strategy: [Eddy, 2023] Space Elevator infrastructures will be compatible with rocket launch infrastructures and complementary in mission support as a member of a Dual Space Access Architecture. One concept circulating is that rockets will move people through radiation belts rapidly while they also support all things Low Earth Orbit. Space Elevators will handle efficient logistic support and move massive amounts to GEO and beyond. In addition, these Space Elevators will enable massive and rapid movement to the Moon and Mars in unmatched trip durations while releasing every day for mission success.

- Operational Distribution: The growth of Space Elevator infrastructures, in the form of Galactic Harbours, will follow a natural pattern. The competition from the commercial owners will be real as well as ownership by/with countries over the years will ensure much needed capabilities. Figure 1.7.1 illustrates this concept of distribution around the equator.

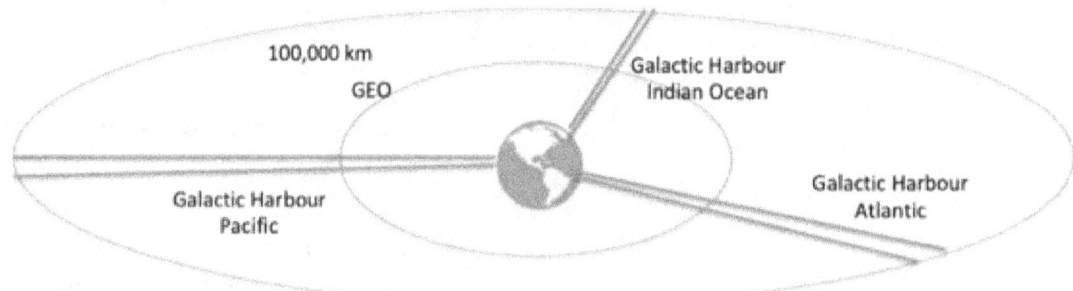

Figure 1.7.1: Estimated Distribution of Galactic Harbours (2038) (ISEC Image)

1.8 Growth of Space Elevators

Over the last 20 years, the ISEC, the Japanese Space Elevator Association (JSEA), and the Obayashi Corporation's expectations of Space Elevator capabilities have come together as an estimate of an IOC in 2036 and FOC in 2040. With the growth of multiple Galactic Harbours towards three around the globe with six Space Elevators, the estimate follows the next two tables: Growth from IOC (until year 12) to FOC (after year 12) and the mass movement are shown in Figure 1.8.1.

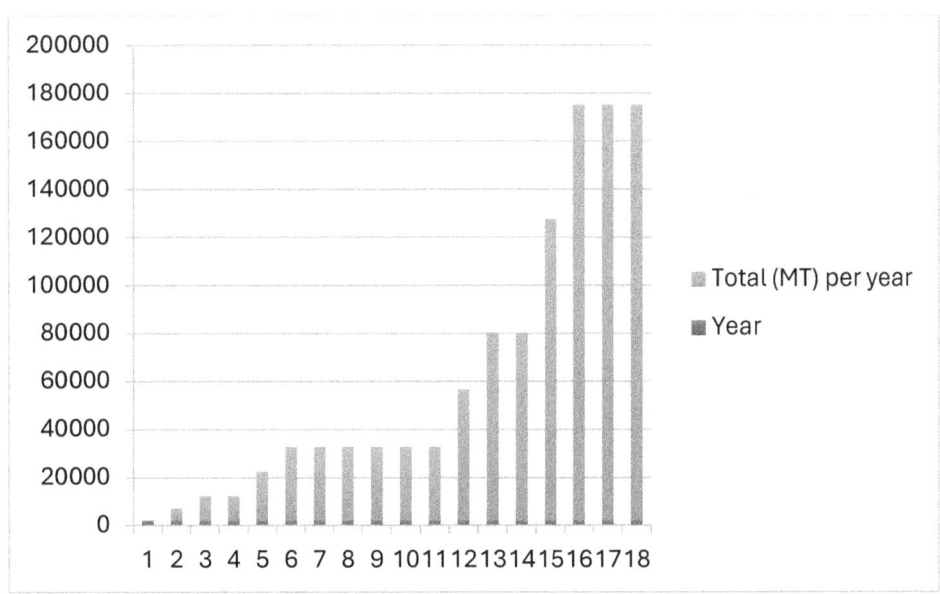

Figure 1.8.1: Projected Mass delivery to GEO
and beyond after IOC(2036) (ISEC Image)

1.9 Baseline Infrastructures:

As shown in the table, the ISEC baseline for Space Elevators, and by extension Galactic Harbours, includes the following characteristics: 100,000 km tether with an Apex Anchor at that altitude, IOC payload capability, an initial space elevator in mid-Pacific (2035) with the first Galactic Harbour growing out of those two tethers (space elevators) by 2036. Each of the major segments (HQ/POC, Apex Anchor, Galactic Harbour, and tether) will be supporting missions with one lift-off a day of 14 MT. The GEO terminus and other points along the tether will be prepared for the activities we propose. [Swan, 2020b]

This leads to the movement of mass with the breakout between IOC and FOC reflecting the maturity of the technology and the desire to have more capability. This leads to a chart showing delivery growth in metric tonnes after the initial Space Elevator IOC. [Swan, 2020b]

1.10 ISEC Apex Anchor Study Starting Point

This study has been authorized by ISEC to illustrate the tremendously unique characteristics of Space Elevator Apex Anchors and compare their future missions that could leverage those revolutionary strengths. The unique capability to operate above gravity well and enable on-time logistics within the CisLunar and Martian regions will become critical for multiple missions of the future. Operating above the gravity well [essentially zero gravity force, with slight acceleration away from the Earth], providing a location to operate as a full-service transportation node, and the ability to release the resulting satellites at tremendous velocities will open up the solar system to all.

We will start with the conclusions from the ISEC 18-month study, produced in 2017, entitled "Design Considerations for the Space Elevator Apex Anchor and GEO Node." [Fitzgerald, 2017] The focus of the study was to analyze the needs at those two locations and how they would be developed during the Space Elevator construction phases. The conclusions were straightforward: "The deployment and continued stability of the tether are the primary function of the Apex Anchor until IOC. This translates to:
- A reel in/reel out (or climb up/climb down) capability,
- A capability to fire thrusters (magnitude and direction) as directed by HQ/POC, and
- Support to customers who leverage the strength of the endpoint of this space transportation infrastructure.

The basic mass buildup for the Apex Anchor will initially be from spent climbers and derelict GEO satellites." [Fitzgerald, 2017] This study was instrumental in establishing the Apex Anchor as a destination on the Space Elevator with great velocity and potential to accomplish many missions not expanded upon. It was obvious from the start that having velocity great enough to release space systems past Mars without rocket fuel opened up tremendous mission of the future. In addition, the ability to sit "above gravity well" enabled so many new concepts. However, the past study was focused upon the completion of the infrastructure and the roles of each segment within the architecture. The writers of this new study report will be starting from the 2017 report with far greater vision into the future after discussing what the transformational characteristics can achieve when one thinks of the needs of humanity in the future. These strengths will be expanded upon in Chapter 2.

1.11 Major Questions for the Study

During this study, there will be many questions addressed to illustrate the transformational characteristics and remarkable strengths. Some of these could be:

a. What is a Full-Service Transportation Node?
b. What is the Apex Anchor Depot Concept?
c. What is a Logistics Depot?
d. What are the principal missions?
e. What are the basic design characteristics of Apex Anchors?
f. How shall the design be driven by potential customers?

g. Differentiate the strengths and weaknesses of departing from the surface of the Earth with building capability and releasing from Apex Anchor?
h. What is the generic footprint on an Apex Anchor platform?
i. Develop for human transport?

1.12 Chapter Breakout

Chapter 1: Introduction and Status of Space Elevators
This will include the characteristics of SEs, introduction to the Galactic Harbour, the projected schedules, the projected mass movement, and conclusion that we are closer than you think –

Chapter 2: Characteristics of Apex Anchor
This chapter will introduce the concept of full-service transportation nodes with multiple special characteristics with emphasis on assembly at top of gravity well. There will be an emphasis on the capabilities of a completed Full Operational Capability with mature operations and sizable loads at 100,000 km altitude enabling full-service transportation nodes.

Chapter 3: Velocity Enhancements
Start with, explaining the velocity characteristics of 100,000 km and 163,000 km. [with table of various altitude release velocities] – then explain the range of velocities that are possible with appropriate expansion of the design of the Apex Anchor region.

Chapter 4: Apex Anchor Missions
This chapter will summarize the capabilities of the full-service transportation node by explaining potential missions from the Apex Anchor. Full Service at the transportation "train like station" will enable missions not even thought of so far.

Chapter 5: Apex Anchor Mission Examples
The strengths of the Apex Anchor enable remarkable missions to the planets in our solar system with extreme velocities and daily departures. A few of the unique missions will be examined.

Chapter 6: Study Summary, Conclusions and Recommendations
This chapter summarizes the Apex Anchor Study and provides recommendations as to future endeavors.

Chapter 2: Apex Anchor as a Full-Service Transportation Node

2.0 Introduction

"Now that we have decided to go to the Moon and on to Mars in a combined international, governmental, and commercial effort of great magnitude, we need to expand our vision of 'how to.' It would seem that the establishment of a Full-Service Transportation Node with an ability to assemble, repair, build and store at a location essentially beyond gravity would open up Human movement beyond our planet." [Swan, 2022] This chapter will discuss the strengths and weaknesses of the Apex Anchor as a significant portion of a permanent transformational space transportation infrastructure. The multiple customer destinations, complexity of releases, magnitude of each transition to transit, and daily launches means humanity can enable monumental "dreams." Expanding space access architectures to include Space Elevators, within a Dual Space Access infrastructure with advanced rockets, will enable a robust movement off-planet.

This study has been authorized by ISEC to illustrate the tremendous strengths of Space Elevators and compare them to the traditional approach of starting from the bottom of the gravity well. The unique capability to operate above gravity well and enable on-time logistics within the CisLunar and Martian regions will become critical for multiple missions. This study's objective is to present the tremendous advantages over rockets by supplying locations [in this case the Apex Anchor] where daily, high speed, releases can reach the Moon in 14 hours and Mars as rapidly as 61 days. These advantages are explained as characteristics or strengths and will provide remarkable capabilities not achievable while depending on the Tsiolkovsky rocket equation. Indeed, Space Elevators will be the transportation story of the 21st century. The key here is that daily, environmentally friendly, routine, inexpensive, efficient delivery, and massive movement of payloads to GEO and beyond will allow high speed travel throughout our solar system. This study will initiate discussions about strengths the space elevators such as supplying: factories, mining operations, distributing CISlunar and solar system satellites.
The reality is when humanity decides to conduct off planet activities, there will be a tremendous need for logistics support, movement of manufactured goods as well as transporting people [especially at low cost and routinely/daily]. The question on ISEC's table is: how can the strengths of space elevators' new and unique capabilities providing a "train station" at 100,000 km altitude enable missions of all types, while having little or no environmental effect on our planet? Inside the train station would be all the functions routinely found on Earth, such as storage – assembly of systems – manufacturing – repair – refueling – daily release – and high speed, with massive payloads, releases.

The proposed question is how many dreams can come true when you assemble large space systems at the Apex Anchor and release them with tremendous velocity? By raising large mass from the Earth's surface, you gain the phenomenal capabilities of Earth manufacturing. By then lifting them in tether climbers at massive rates per day enables future settlements to have what they need delivered efficiently and "on-time" – similar to an Earth based train station. In addition, by assembling and repairing at the Apex Anchor, the customer gains the option of even larger [than one tether climber load] space systems released daily towards their needs. "Previous studies conducted by the International Space Elevator Consortium (and the International Academy of Astronautics) have shown that the initial operational capabilities of 30,000 tonnes to GEO and beyond per year will be a phenomenal jump from rocket delivery. In addition, the estimate of full operational capacity is 170,000 tonnes per year to GEO and beyond. When you realize that the restrictions of rocket launches are well understood, and when you combine that with space elevators, which essentially beat the gravity well, the capabilities of space elevators become obvious and.

- Enable Space Solar Power while supporting the Paris Accords
- Lift payloads as the Green Road to Space, helping to save our atmosphere
- Improve life on Earth with major accomplishments, in space
- Enable early completion of massive projects, such as lunar villages • shorten the time for delivery of 1,000,000 tonnes to Mars, and
- Enable early development of an L-5 settlement with millions of inhabitants." [Swan, 2022]

2.1 Modern-Day Space Elevator

Discussions about Modern-Day Space Elevators impact our future movement beyond LEO and its transformational aspects at a basic level. As permanent space transportation infrastructure, they will lead to daily, routine, unmatched efficiencies, environmentally friendly, massive movement of cargo and inexpensive departures towards mission destinations throughout the solar system. ISEC believes that the Modern-Day Space Elevator is closer than you think and will fascinate the space community with its remarkable capabilities for humanity's future missions. These new capabilities are revolutionary in approach, but evolutionary in scope. Its major characteristics today are:

- Space Elevators are entering engineering development,
- Space Elevators are the Green Road to Space,
- Space Elevators are a major portion of the future Dual Space Access Architecture teaming with advanced rockets, and
- Space Elevators are a permanent transportation infrastructure moving massive cargo daily, efficiently, routinely, and safely.

The Modern-Day Space Elevator received this nomenclature from the realization that the material for the 100,000 km tether should be available in time for the construction of the full-length initial deployment. [Wright, 2023] As the development of the Modern-Day Space Elevator can be initiated, the future becomes more in line with the dreamers such as Glaser and Mankins [Solar Power], the National Space Society's vision of an L-5 Settlement, and Musk looking towards Mars development. When the Galactic Harbour architecture becomes operational, the transportation aspects of access to space will be focusing upon the similarities of railroad operations with a permanent set of "tracks" vertical to GEO, the Moon and beyond. These transportation characteristics will lead access to space towards similarities of the change from small boats crossing a river to a permanent bridge carrying immensely more cargo routinely, safely, inexpensively with far greater efficiencies. Indeed, Full-Service Transportation Nodes will become operational at such locations as GEO and the Apex Anchor.

2.2 Permanent Space Access Infrastructure

In Space Elevator concept, climbers are the vehicles while tethers, Earth Port with several termini and operational platforms, GEO construction - repair garages/stations, and Apex Anchor are permanent, reliable space infrastructures. These permanent transportation infrastructures define the future space superhighway's green road to space with collaborating and complementary permanent infrastructures such as space stations in low Earth orbit. These commercially developed Galactic Harbours will have competition between themselves and they will be complementary with advanced rockets within a Dual Space Access Architecture (see latest ISEC study researching topic) [Eddy 2023]. In addition, as climbers are raised by electricity, the system can be called the Green Road to Space [Eddy 2021]ensuring that lifting payloads to GEO and beyond will not impact our fragile environment while delivering massive cargo such as Space Solar Power satellites. A recent insight references the economics of permanent space access infrastructures by two authors at the International Astronautical Congress IAC-21.

> "This paper analyzes the economics of Space Elevators as infrastructure and a platform, utilizing relevant historical examples, such as the standardization of shipping containers, the transcontinental railroad, and the Panama Canal to explore its economic value and developmental impact. Infrastructure, at its core, provides value through the reduction of transaction costs. Therefore, trying to close a business case for infrastructure by charging high transaction costs is a doomed venture. However, expanding the picture to view the impact on the economy from increased access to value and more efficient markets through lower transaction costs and infrastructure becomes a very lucrative, stable, and reliable investment. Cost per kilogram is the language of rockets -- strategic investment, ubiquitous access, and uninterrupted exchange of resources are the staples of Space Elevators." [Barry, 2021]

Placing the new economic "value proposition" approach towards development and operations empowers the statement: We must build space elevators as soon as possible. The time is NOW. The Green Road to Space must be initiated as soon as possible.

2.3 ISEC Study Starting Point

The unique capability to operate above gravity well and enable on-time logistics within the CisLunar and Martian regions will become critical for multiple missions. When Galactic Harbours are operational, the similarities to railroad operations will be remarkable. The following set of seven permanent transportation infrastructures characteristics will have the ability to enable customers' missions. The transformation of space access will be similar to moving from small boats crossing a large river to a permanent infrastructure (a bridge) moving traffic daily, routinely, safely, inexpensively, and with little environmental impact. Permanent space transportation infrastructures (space elevators) with Full-Service Transportation nodes (at GEO and Apex Anchor) will enable missions by leveraging their strengths.

Figure 2.3.1: Significant Strengths of Space Elevators

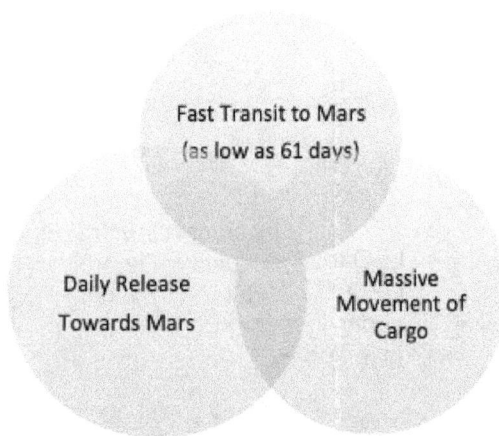

Essentially, their strengths are permanent infrastructures that lead to daily, routine, environmentally friendly and inexpensive departures towards mission destinations (see Figure 2.3.1). This inherent capability results from their "permanent" characteristics leading to a bridge into space. Once Space Elevators have been installed and upgraded to their initial capability, they will be there for a very long time, similar to roads, bridges, and railroads tracks. Rising from the surface of the ocean to Apex Anchors (100,000 km as a starting concept) is accomplished with external power - such as solar energy. The permanent transportation infrastructures of space elevators will enable missions by leveraging their strengths:

- Daily, routinely, safely, efficiently, inexpensively: The preliminary design of space elevators has shown there will be one climber initiating its departure from each Earth Port at sunrise each day with a seven-day travel time to the GEO region and then on to the Apex Anchor. This enables logisticians to send a 100-tonne climber (79 tonnes are payload) up a space elevator daily, routinely, safely, and inexpensively (see Figure 2.3.2).
- Assembly at the Top of the Gravity Well: The idea is simple – raise payloads with electricity to 100,000 km altitude and then assemble them in a robotic factory at the Apex Anchor. This leads to an operational capability which will release any size spacecraft, with appropriate rocket motors, to reach any planet in any inclination daily and safely – all while being environmentally safe.
- Unmatched Delivery Statistics: The Space Elevator concept has an unmatched efficiency strength for moving mass to space on a permanent transportation space system: 70% vs. 1% to the Moon. Essentially, rockets deliver 2% of the launch mass to GEO altitude and 0.5% of launch mass to the surface of the Moon (or Mars). The efficiency of space elevators is in the range of 70% of liftoff mass arrives at its destination (with the remaining 30% surviving in the form of a reusable tether climber).

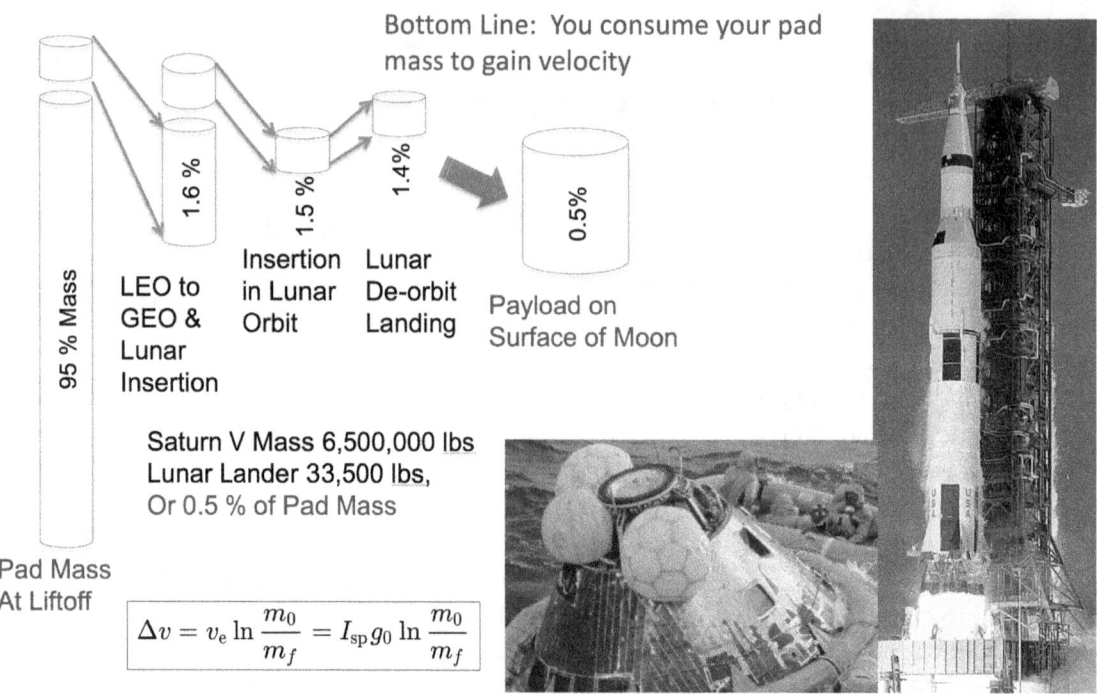

Figure 2.3.2: Rockets

- Massive movement (Initial Operational Capability (IOC) at 30,000 tonnes/yr and Full Operational Capability (FOC) 170,000 tonnes/yr) [Swan, 2020]: As noted earlier, the ability to move 30,000 tonnes during initial operations will be revolutionary when compared to our rocket history. When considering that Space Elevator complexes will grow to a capacity of 170,000 tonnes per year, the belief is born that any mission can be fulfilled. There will be no limitations on the needs of customers as Space Elevators can deliver every day of the year to any location at GEO and/or beyond. A chart showing the natural growth of Space Elevators' capacity is shown in Figure 2.3.3.

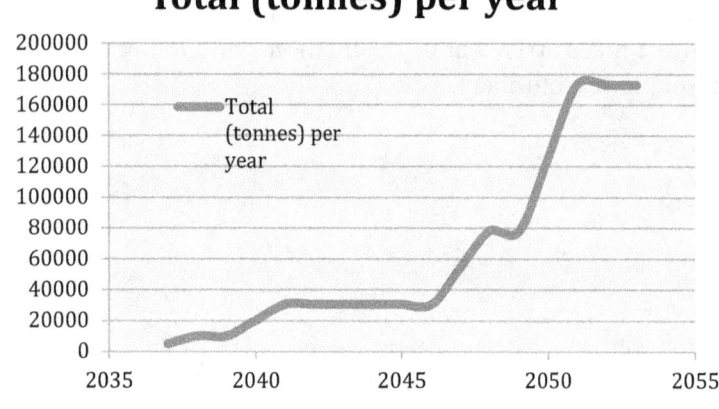

Figure 2.3.3: Massive Cargo Movement by Space Elevators [Swan, 2021][1]

- As a Green Road to Space, it ensures environmentally neutral operations: Space Elevators are environmentally neutral as they raise tether climbers using electrical energy from the sun. This enables climbers to transfer through the atmosphere without impacting it

---

[1] MT is metric tonnes.

with combustion by-products. As the process is continuous with a single tether climber, there is no residual hardware left before its destination (space debris), especially in the Low Earth Orbit region.

- Reduction of the need for Rocket Fairing Design limitations: The beauty of a permanent transportation infrastructure, such as Space Elevators, is that cargo can move across distances efficiently, effectively, and with little effect on the cargo. This intermodal transportation technique is called containerizing. The lack of rockets' shakes, rattle, and roll will simplify the design of satellites as well as storage and transportation requirements for transit to GEO and beyond (see Figure 2.4)

- Unmatched High velocity [Delta V] (starting at 7.76 km/sec at 100,000 altitude enables rapid transits to the Moon, Mars and beyond): This strength results from the increase in velocity as tether climbers raise their payloads. The rotation of the Earth with a long "arm" instills tremendous velocity as altitude is gained. At GEO, the velocity matches the orbit and then it increases rapidly as the altitude is increased. At the 100,000 km altitude of the Apex Anchor, the velocity is 7.76 km/sec, or enough to go beyond Mars with no extra propellent or reach Mars in as few as 61 days. [see Figure 2.3.4]

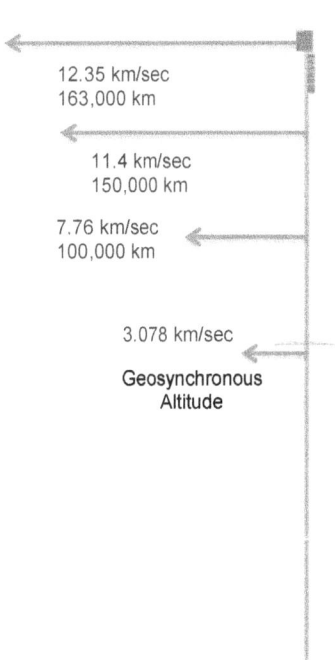

Figure 2.3.4: Release Geometries

- Transforming the economics towards an infrastructure with access to more valuable, lucrative, stable, and reliable investments [Barry, 2021]. This transformation results from the economic impacts of permanent space access infrastructures. From the beginning of this century, the Space Elevator Community has played the rocket game – trying to show that it will be the least expensive transportation infrastructure. We CAN "beat" this low cost at so many levels; but the discussion needs to be raised to another level by actually explaining that Space Elevators reach across economic growth arenas of enterprises across the solar system.

2.3.1 Growth of Space Elevators

Space Elevators expectations – between the ISEC and the Japanese Space Elevator Association (JSEA) – have been estimated to have an IOC in 2036 and FOC in 2040. The growth of multiple Galactic Harbours is expected to be supported by other countries (as a launch capability) and by industry (as a center for massive enterprise growth). The IOC Space Elevators are at 30,000 tonnes per year with six tethers while the FOC Space Elevators are at 173,000 tonnes per year with six tethers.

Table 2.3.1.1: Growth of Space Elevators Initial to Full Operations

Initial Operational Capability (IOC)	First Space Elevator (IOC)	2036
	First Galactic Harbour (Mid-Pacific)	2036
	Second Galactic Harbour (Indian Ocean)	2037
	Third Galactic Harbour (Mid-Atlantic)	2038
Full Operational Capability (FOC)	First Space Elevator (FOC)	2040
	First Galactic Harbour (Mid-Pacific)	2041
	Second Galactic Harbour (Indian Ocean)	2041
	Third Galactic Harbour (Mid-Atlantic)	2042

2.3.2 Compatible Dual Space Access Infrastructures

Space Elevator infrastructures will be compatible with rocket launch infrastructures and complementary in mission support as a member of a Dual Space Access Architecture (see Figure 2.3.2.1). One concept circulating is that rockets will handle people for rapid movement through radiation belts while early Galactic Harbours will focus on logistic support. [Eddy 2023]. In this Architecture, rockets will open up the Moon and Mars and initiate such dreams as Space Solar Power. Then Space Elevator Transportation Systems will leverage their ability to do heavy lifting while their strengths will supply and grow the dreams of many.

Figure 2.3.2.1: Space Elevator Infrastructure

Chapter 3: Velocity Enhancements

3.1 Introduction

Chapter 3 deals with the opportunities for achieving a greater choice of velocities, leading to more control over both speed and direction, enabling even shorter flight times to the planets and great flexibility over the choice of launch dates. It is also possible to contemplate flights to the Oort cloud and beyond.

In the standard model of the space elevator, the tether extends 100,000km from the earth's surface. As it rotates with the earth, it has the same angular velocity, and it follows that the velocity at the apex anchor is 7.7km/s in the direction of a circular orbit in the plane of the equator relative to the Earth's center. A spacecraft released here will continue in the Earth's equatorial plane while in Earth's gravity well, but when it leaves the Hill Sphere (passing the point at which the sun's gravity dominates), it will proceed in orbit about the Sun and will thus be able to reach Mars and other destinations in the solar system with fuel needed only for course corrections and deceleration.

One way to increase the velocity is to lengthen the tether. For example, lengthening it to 120,000km results in a relative velocity of 9.2km/s in the same direction; at 150,000km it is 11.4km/s and at 163,000 Km the velocity 12.35Km/s is sufficient to escape the Earth's gravity.. The tether taper equation means that any length of tether is theoretically possible, but at extreme altitudes the required Anchor mass will fall too low for stability control and other Anchor functions (unless the tether maximum thickness at GEO is increased).

Figure .1 shows how the tether and anchor masses vary with altitude for a fixed GEO tether thickness.

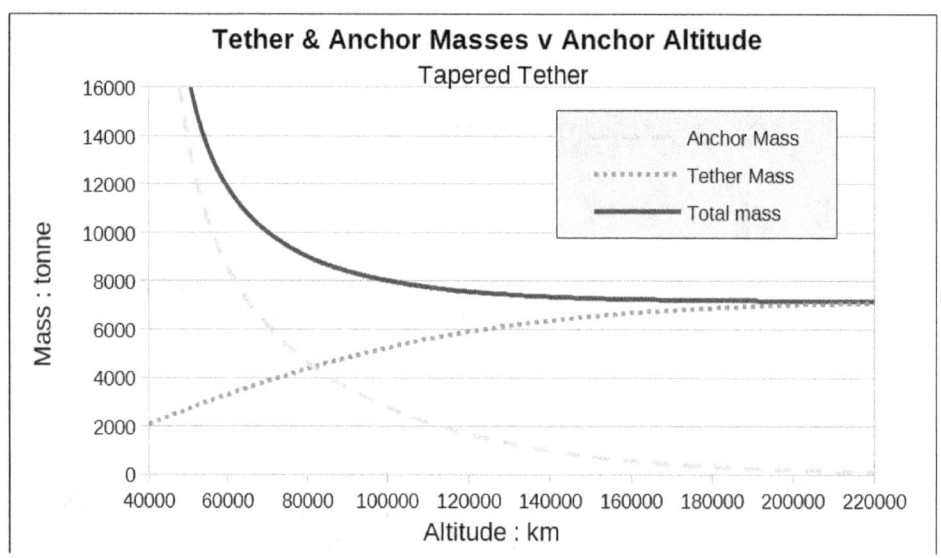

Figure 3.1.1 Tether and Anchor Masses v. Altitude for a fixed tether area at GEO

A tether as long as 200,000km is possible if we place a large mass near the earth to balance the extra length of tether. Of course, the tether must be made thicker. Then the apex anchor will travel at 15km/s. If the reference tether cross-section at GEO is increased then the required Anchor masses would increase pro rata, but system stability would require a higher Earth Port retention force or perhaps some substantial mass suspended below GEO: for example, this large mass could be a 1000-tonne hotel at 1580km altitude.

The advantage of increasing the Anchor altitude is commonly described as yielding a higher circumferential release velocity, but this benefit can also be expressed as a higher specific orbital energy (SOE). The SOE value considers the potential gravitational energy of a body and remains fixed for a body in a free orbit, thus being a better measure of the orbital behavior of a spacecraft released for an interplanetary journey. Figure 3.1.2 shows the SOE for a body at altitudes between the Earth's surface, compared with the SOE of a satellite in a circular orbit.

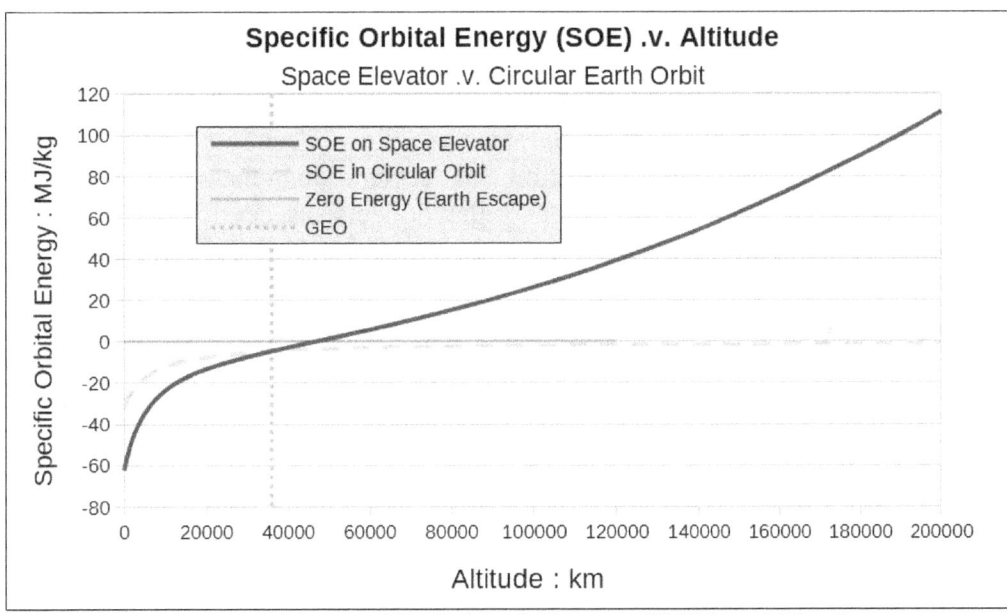

Figure 3.1.2 Specific Orbital Energy (SOE) v Altitude

Peet [Peet, 2021] proposes to increase the release velocity further and give more flexibility over its direction by taking advantage of the centrifugal force along the tether to accelerate a spacecraft radially. Beyond GEO, the outward force on the tether and on any craft traveling on it exceeds the pull of gravity, and so the craft will naturally accelerate away from Earth. At 100,000km from the earth's surface, it will reach an outward speed along the tether of 6.2km/s. Then the net release velocity will be $\sqrt{7.7^2 + 6.2^2} = 9.9$km/s at an angle $\tan^{-1} = 51°$ to the tether in the equatorial plane. This 9.9 km/s net release velocity can be shown to yield a specific orbital energy like that of a body released from an altitude of 150,000 km with zero radial velocity.

If we lengthen the tether to 150,000km, the free ascent speed along it will be 10km/s. Then the net velocity will be $\sqrt{11.4^2 + 10^2} = 15.2$km/s at an angle $\tan^{-1}\frac{11.4}{10.0} = 48.7°$ to the tether in the equatorial plane.

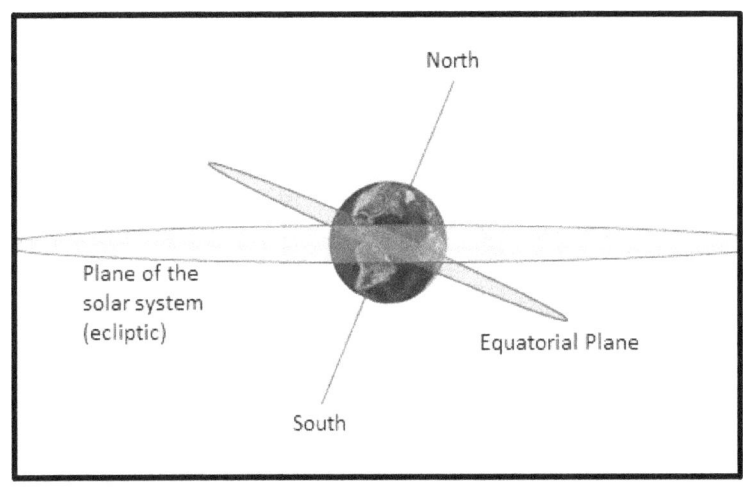

Figure 3.1.3 The plane of the planets' orbits and the plane of the space elevator

Professor Peet points out significant limitations to this scheme, because there are just two occasions per day when the apex anchor will be aligned with the plane of the solar system which is known as the ecliptic (Figure 1.3) and these occasions are not necessarily those needed to transfer to a particular planet. Furthermore, he shows by a lengthy mathematical argument that, over the next century, launching is limited to the solstices unless a tether 150,000km long is used. Of course, there is always the option of launching a rocket-powered spacecraft to achieve the desired trajectory, but this treatment concentrates on trajectories that do not need fuel apart from minor course correction and deceleration at the destination. For these reasons Peet introduces a ramp at the apex anchor that deflects spacecraft to the desired direction (Figure 3.1.4). The ramp rotates slowly about the tether and thus introduces another degree of freedom that gives good directional control.

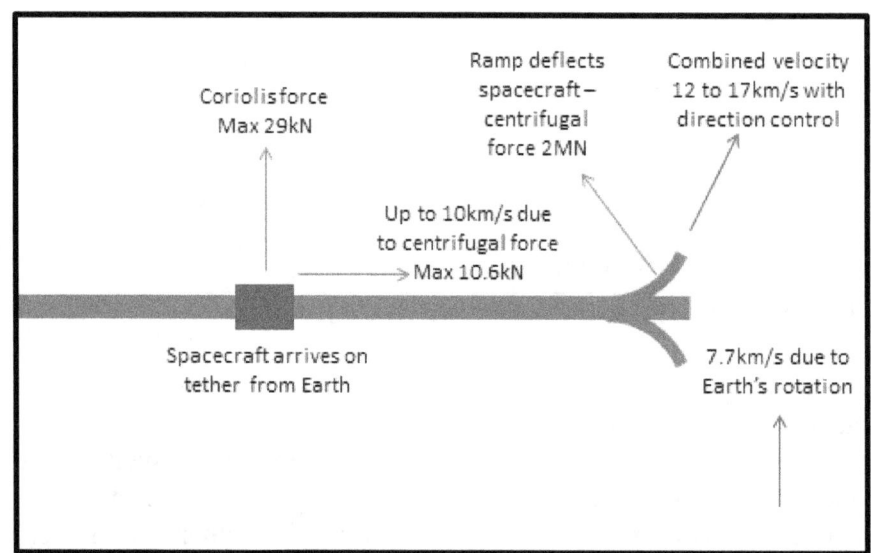

Figure 3.1.4 Spacecraft approaches a ramp that deflects it to the required direction

3.2 Critique

Allowing spacecraft to reach speeds up to 10km/s along the tether raises significant concerns about the feasibility of the climber/tether interface (given that wheels cannot operate at such speeds) and damage to the tether. Not only will spacecraft travel along the tether, but they will also rely on the tether to accelerate them in the equatorial plane to achieve the circumferential speed of 7.7km/s or more, exacerbating the interface feasibility concerns: near the ramp, this Coriolis force will reach 29kN on a 20-tonne craft. On the ramp itself, an additional force is caused by the deflection, and this can be as high as 2MN. Spacecraft would thus need to be strong enough to withstand an acceleration of $100m/s^2$, which is about $10g$, where g is the acceleration due to gravity on the earth's surface. The ramp will also be deflected, and so a dummy mass would need to be applied to the opposite ramp simultaneously or soon afterwards to stabilize it.

The addition of the rotating ramp permits spacecraft to travel to all solar system destinations in suitable time without the use of fuel except for minor course correction and deceleration (see Table 2.1).

Table 3.2.1 Travel times to the planets

	Fastest time per synodic period (days)	Longest time (days)	Synodic period (years)
Mars	60	800	2.1
Jupiter	450	850	1.1
Saturn	800	1165	1.0
Uranus	2000	2365	1.0
Neptune	4000	4365	1.0

The synodic period is the time it takes for a celestial body to return to the same position with respect to the sun as seen from Earth.

3.3 The Celestial Sphere

Once upon a time, it was thought that the earth was surrounded by crystal spheres. Although that theory has long since been debunked, astronomers still imagine the earth surrounded by a sphere to determine the positions of the stars and other celestial bodies. Two angles are sufficient to define any direction out from Earth, by analogy with longitude and latitude. The details are different, and the astronomical measurements are right ascension and declination, but the essential point is that two angles in two planes are necessary and sufficient to define the direction from Earth to any celestial body.

The space elevator tether rotates in the equatorial plane. By suitable timing, a spacecraft traveling along the tether will leave it at the apex anchor in the desired direction in that plane. Its encounter with the rotating ramp will be timed to cause it to be transferred out of the equatorial plane at the required angle.

3.4 Slingshot

Peet also describes a slingshot method in which a large mass is used to accelerate a small mass. The idea is to have the two masses connected by an elastic tether and to send them on opposite sides of the ramp (Figure 3.4.1).

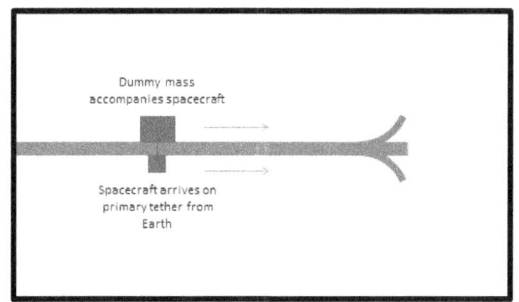

Figure 3.4.1 Dummy mass travels with spacecraft

The two masses thus leave the ramp traveling in opposite directions. (Figure 3.4.2) When the connecting tether becomes taut it will function as a spring, causing the large mass to pull the small mass towards it. At the end of the spring-induced acceleration the smaller spacecraft can be released at a much higher velocity. (Figure 3.4.3)

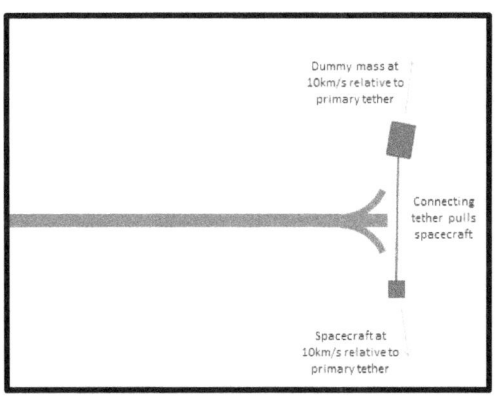

Figure 3.4.2 Spacecraft travels away from dummy mass until connecting tether becomes taut

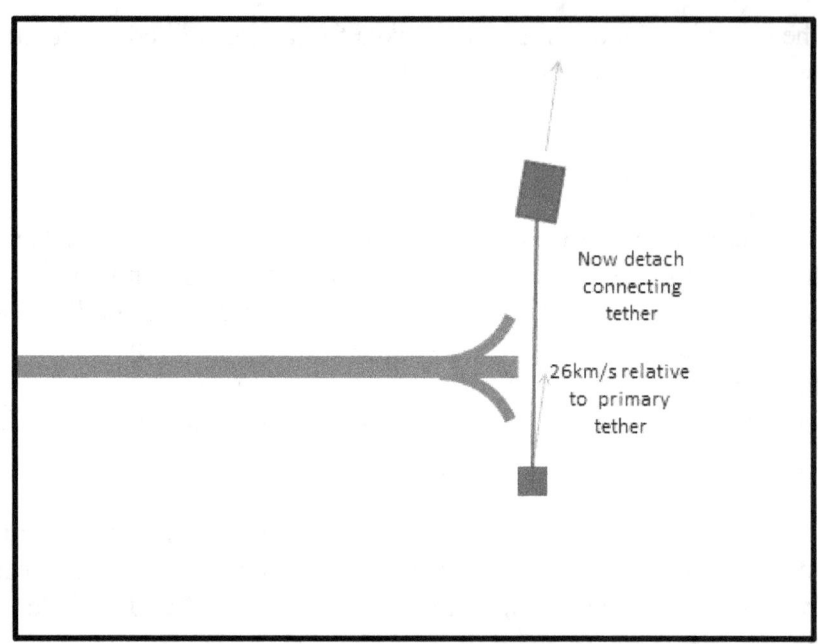

Figure 3.4.3 Dummy mass pulls spacecraft at much higher velocity

Let the masses be m_1 and m_2 with $m_2 > m_1$ and their velocities $-v$ and v on emerging from the ramp. The connecting tether changes the velocities to v_1 and v_2. Conservation of momentum implies:

$$m_1 v_1 + m_2 v_2 = (m_2 - m_1)v$$

$$\therefore v_2 = \frac{(m_2 - m_1)v - m_1 v_1}{m_2}$$

$$\therefore v_2^2 = \frac{[(m_2 - m_1)v - m_1 v_1]^2}{m_2^2}$$

Conservation of energy implies:

$$m_1 v_1^2 + m_2 v_2^2 = (m_2 + m_1)v^2$$

$$\therefore v_2^2 = \frac{(m_2 + m_1)v^2 - m_1 v_1^2}{m_2}$$

Hence

$$\frac{(m_2 + m_1)v^2 - m_1 v_1^2}{m_2} = \frac{[(m_2 - m_1)v - m_1 v_1]^2}{m_2^2} \tag{3.1}$$

The solution is:

$$v_1 = \frac{3m_2 - m_1}{m_1 + m_2} v$$

The derivation is in the section 3.5 **Calculations**. If $m_2 = 10 m_1$ then $v_1 = 2.64v$. For example, if the spacecraft is two tonnes and the slingshot mass is twenty tonnes and they emerge from the ramp at 10km/s in opposite directions, the spacecraft reaches 26km/s.

Assuming strong enough connecting tethers, it is possible to apply this principle more than once and get further increases in spacecraft velocity but with smaller and smaller spacecraft. Adding one more stage would mean that the spacecraft is 200kg in mass. A similar calculation reveals that the exit velocity of this spacecraft would be 56km/s. If we align the ramp with the earth's rotation about the sun, which is 29.78km/s the net velocity relative to the sun will be 85km/s – fast enough to exit the solar system at more than 40km/s.

Of course, there are serious engineering questions to be answered, but the possibilities are endless.

3.5 Calculations

The calculations mentioned in the text are given here.

From equation (3.1)

$$m_2[(m_2 + m_1)v^2 - m_1v_1^2] = [(m_2 - m_1)v - m_1v_1]^2$$

$$\therefore m_2[(m_2 + m_1)v^2 - m_1v_1^2] = (m_2 - m_1)^2 v^2 - 2m_1v_1(m_2 - m_1)v + m_1^2 v_1^2$$

$$\therefore 0 = m_1(m_1 + m_2)v_1^2 - 2m_1(m_2 - m_1)vv_1 + [m_2^2 - m_2^2 - 3m_1m_2 + m_1^2]v^2$$

$$\therefore 0 = m_1(m_1 + m_2)v_1^2 - 2m_1(m_2 - m_1)vv_1 + m_1(m_1 - 3m_2)v^2 = 0$$

$$\therefore 0 = (m_1 + m_2)v_1^2 - 2(m_2 - m_1)vv_1 + (m_1 - 3m_2)v^2 = 0$$

$$\therefore v_1 = \frac{(m_2 - m_1) \pm \sqrt{(m_2 - m_1)^2 - (m_1 + m_2)(m_1 - 3m_2)}}{(m_1 + m_2)} v$$

$$\frac{(m_2 - m_1) \pm \sqrt{m_2^2 - 2m_2m_1 + m_1^2 - m_1^2 + 2m_2m_1 + 3m_2^2}}{(m_1 + m_2)} v$$

$$\therefore v_1 = \frac{(m_2 - m_1) \pm \sqrt{4m_2^2}}{m_1 + m_2} v$$

$$\therefore v_1 = \frac{(m_2 - m_1) \pm 2m_2}{m_1 + m_2} v = \frac{3m_2 - m_1}{m_1 + m_2} v \lor -v$$

The result $v_1 = -v$ is the trivial case where the spacecraft carries on as before. The interesting case is $v_1 = \frac{3m_2 - m_1}{m_1 + m_2} v$.

The calculations for a three-body case are as follows.

Let this third object be the spacecraft with a mass of m_s, an initial velocity of $-v$ and an eventual velocity of v_s. Then the conservation of momentum equation is

$$m_1 v_3 + m_s v_s = m_1 v_1 - m_s v$$

$$\therefore v_3^2 = \frac{[m_1 v_1 - m_s v - m_s v_s]^2}{m_1^2}$$

The conservation of energy equation is:

$$m_1 v_3{}^2 + m_s v_s{}^2 = m_1 v_1{}^2 + m_s v^2$$

$$\therefore v_3{}^2 = \frac{m_1 v_1{}^2 + m_s v^2 - m_s v_s{}^2}{m_1}$$

Hence

$$\frac{m_1 v_1{}^2 + m_s v^2 - m_s v_s{}^2}{m_1} = \frac{[m_1 v_1 - m_s v - m_s v_s]^2}{m_1{}^2}$$

$$\therefore m_1 [m_1 v_1{}^2 + m_s v^2 - m_s v_s{}^2] = [m_1 v_1 - m_s v - m_s v_s]^2$$

$$\therefore m_1{}^2 v_1{}^2 + m_1 m_s v^2 - m_1 m_s v_s{}^2$$
$$= m_1{}^2 v_1{}^2 + m_s{}^2 v^2 + m_s{}^2 v_s{}^2 - 2 m_1 m_s v_1 v - 2 m_1 m_s v_1 v_s + 2 m_s{}^2 v v_s$$

$$\therefore 0 = m_s(m_s + m_1) v_s{}^2 + 2(m_s v - m_1 v_1) m_s v_s + (m_s - m_1) m_s v^2 - 2 m_1 m_s v v_1$$

Try some numbers: $m_s = 0.2, m_1 = 2.0, v_1 = 2.64 v$.

$$0.2(2.2) v_s{}^2 + 2\big(0.2 v - 2.0(2.64 v)\big) 0.2 v_s + (0.2 - 2.0) 0.2 v^2 - 2(2.0 * 0.2) 2.64 v^2$$

$$\therefore 0.44 v_s{}^2 + (0.4 - 10.5) 0.2 v v_s - 0.36 v^2 - 2.11 v^2 = 0$$

$$\therefore 0.44 v_s{}^2 - 2.02 v v_s - 2.47 v^2 = 0$$

$$\therefore v_s = \frac{2.02 \pm \sqrt{4.08 + 4 * 0.44 * 2.47}}{0.88} v = \frac{2.02 \pm \sqrt{4.08 + 4 * 0.44 * 2.47}}{0.88} v = \frac{2.02 \pm 2.90}{0.88} v$$

$$\therefore v_s = 5.59 v$$

3.6 Secondary Tethers

Another method of controlling the speed and the direction of spacecraft emerging from the apex anchor is a system of secondary tethers rotating about the primary tether at the apex anchor. [Knapman, 2022]. In this concept the spacecraft is never subject to a g-force greater than one, which is the same as that experienced on Earth due to gravity. Secondary tethers are therefore suitable for delicate payloads, including people eventually. A good safety record with unmanned spacecraft will be needed before human beings are allowed to travel on it, but (unlike the Peet high-speed release scenario) it does mean that spacecraft need not be made very robust, thus reducing weight and cost, and allowing more capacity for useful payloads. It is also the case that the low costs and the chance to launch spacecraft often will allow spacecraft to be built more simply and with less backup and failsafe mechanisms because launches to many destinations can take place daily instead of just once in a decade or two.

When a spacecraft arrives at the apex anchor, it will be directed to one of six secondary tethers. It will travel along it in the same way as it travels along the primary tether using the centrifugal force as the means of propulsion. The primary tether rotates about the earth once a day. The secondary tethers rotate about the apex anchor 13.7 times per day. With a length of 10,000km the speed at the end is 10km/s, and so this is the speed at which a spacecraft will be launched. To this we can add the velocity of the apex anchor, which is 7.6km/s at 100,000km from Earth. Depending on the time that the spacecraft is released, the combined velocity will vary between 17.7km/s and 12.6km/s.

If we allow spacecraft to accelerate along the secondary tether under the centrifugal force due to rotation about the apex anchor, that will yield another 10km/s orthogonal to the peripheral velocity, giving a resultant velocity of 14.1km/s. When this is combined with rotation about the earth, the net velocity will vary between 21.7km/s and 16.1km/s, depending on the time and hence the angle of release. If the primary tether is 150,000km long, the velocity at the apex anchor is 11.4km/s, and the resultant velocity of the spacecraft using a secondary tether range from 25.5km/s to 18.1km/s. If the release is so timed, we can arrange for the spacecraft to be released in the direction of Earth's orbit about the sun. Then the maximum net velocity relative to the sun with a primary tether length of 100,000km is 51km/s, which is enough to leave the solar system with about 10km/s velocity.

3.7 Description

Altogether there are six secondary tethers and six reaction tethers (Figure 3.7.1). There are three outer secondary tethers spaced 120° apart to maintain a balance. When a spacecraft is to be launched from one of the secondary tethers, the other two secondary tethers need to be loaded with ballast to maintain the balance.

The inner secondary tethers are configured like the outer ones. Their purpose is to counteract the tendency of the system to rotate (Newton's second law) as additional rotational energy is added to the secondary tethers while the spacecraft and ballast move along the tethers and, in so doing, increase their speed. There will be a tendency for the secondary tethers to get faster over the course of multiple launch operations, but this problem can be mitigated by alternately using the outer and inner secondary tethers. The inner secondary tethers rotate in the opposite direction to the outer ones.

The reaction tethers are shorter, 1100km instead of 10,000km. They are needed to counteract the gyroscopic effects of rotating about the apex anchor while also rotating about the earth. There is one set for each set of secondary tethers. In each case the reaction tethers rotate in the opposite direction to their secondary tethers.
The reaction tethers are connected to their secondary tethers via a strong connector. It is short, about three meters, and can sustain the bending caused by the gyroscopic effects. Connecting the inner and outer sets of secondary and reaction tethers is an axle 4.3km long. It only must deal with rotational forces and not bending. The spacing is needed to allow for movement caused by the gyroscopic effects, as explained in [Knapman, 2022].

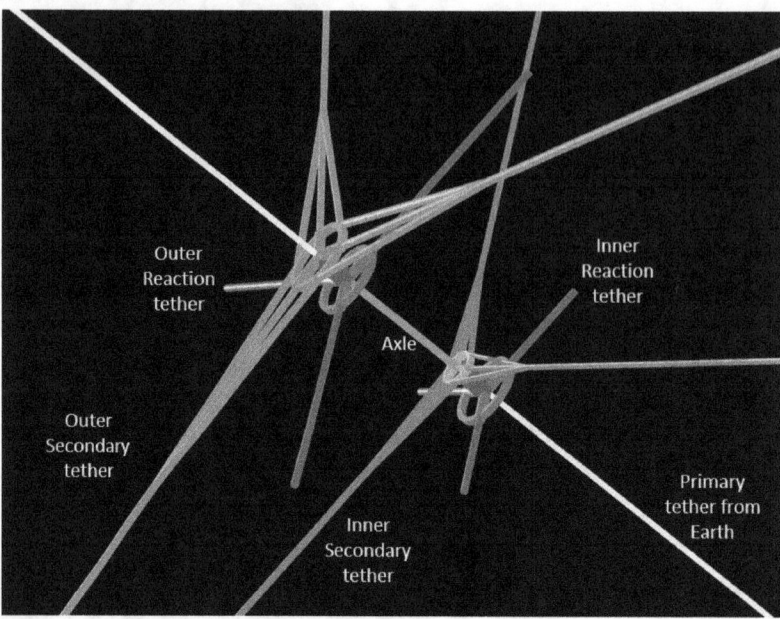

Figure 3.7.1 Arrangement of secondary and reaction tethers

At the center of each set of three tethers is a drive wheel. Not shown are the electric motors and the solar panels to power them. The arrangement of short supplementary tethers supporting each secondary tether is there to ensure effective, stable acceleration as described in [Knapman, 2022] Graphene is assumed as the material to use, which is also the preferred material for the primary tether. Table 3.7.1 shows the outline mass budget for the design described here.

Table 3.7.1 Mass budget for secondary tethers and supporting facilities

	Number	Each (tonnes)	Total (tonnes)
Secondary tethers	6	120	720
Drive wheels	2	20	40
Reaction tethers	6	120	720
Reaction wheels	2	20	40
Electric motors	2	75	150
Solar panels	2	20	40
TOTAL			1710

The journey of a spacecraft along a secondary tether starts with gentle acceleration up to 500km/hr. See Figure 3.7.2. It then takes 20 hours to reach the launch point at the end.

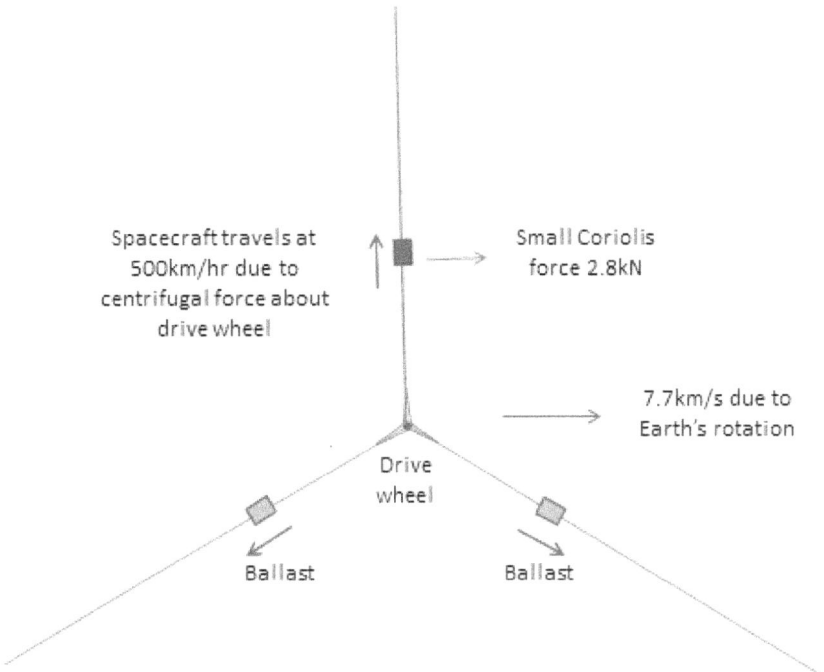

Figure 3.7.2 Gentle ride to tether's end

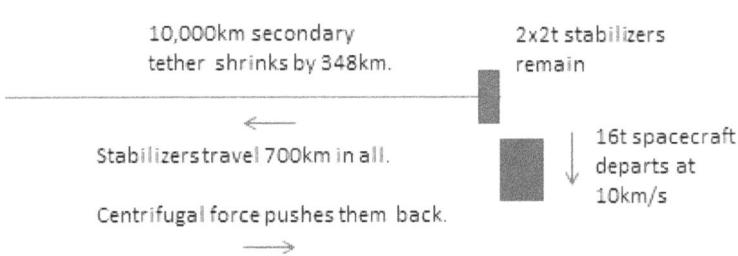

Figure 3.7.3 Method of dealing with recoil

When a spacecraft is released from a secondary tether, there will be a recoil as the secondary tether is suddenly relieved of a considerable amount of tension. To deal with this we use stabilizers, which remain attached. For a spacecraft of sixteen tonnes, two stabilizers of two tonnes each remain attached to the tether. Judicious separation of these two tethers provides a form of active damping, which prevents prolonged oscillation. [Knapman, 2023] Figure 3.7.3 shows the point of departure of the spacecraft, Figure 3.7.4 shows the separation of the two stabilizers, and Figure 3.7.5 has a graph of the motion of the two stabilizers, which rapidly settle down.

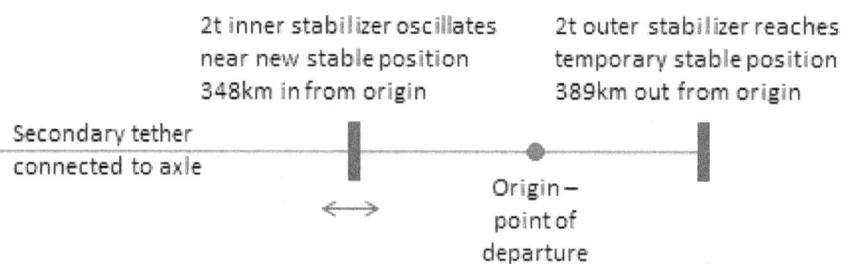

Figure 3.7.4 Separation of stabilizers

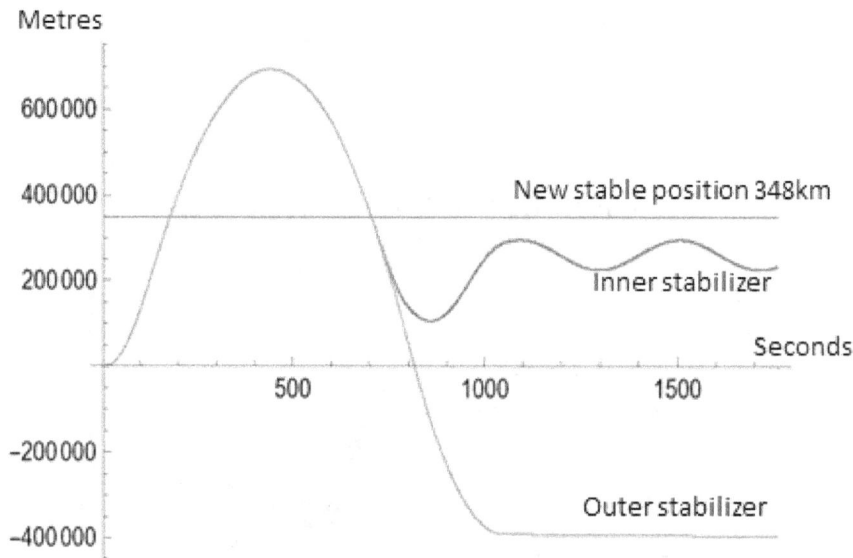

Figure 3.7.5 Motion of stabilizers

International Space Elevator Consortium   *ISEC Position Paper #2024*

Chapter 4: Apex Anchor Missions

4.0 Introduction

It is envisioned, once the Apex Anchor is operational, that a host of missions can be performed. This Chapter provides a short summary of possible missions within the Apex Anchor.

4.1 Apex Anchor Operations Center

Mission: Providing counterweight stability for the space elevator as a large end mass, the Apex Anchor will house a Center much like the bridge on a ship which is tasked to control all operations. It will support all day-to-day operations and host a variety of space missions and operations such as these already being designed.

Figure 4.1.1: Apex Anchor Operations Center

This Photo by Unknown Author is licensed under CC BY-SA

Operations Center for vital mission support. This could include: space domain mission awareness, space warning center, space surveillance and reconnaissance, position and navigation reference point, communications beacon, communications relay station, CISlunar active navigation support, as well as the Earth Defense System.

4.2 Planetary Defense

Mission: Protect the planet from asteroid impacts and other celestial hazards . The Space Elevator Planetary Defense Mission Control will play a pivotal role in managing various activities related to planetary defense, focusing on essential functions such as searching, detecting, tracking, characterizing, planning, and coordinating responses to objects approaching Earth.

Figure 4.2.1: Incoming Asteroid[2]

Humanity's existence on the Earth depends upon Planetary Defense systems and their timely deployment. The authors believe that several asteroids pass the Earth from the Sun side without being identified until after they have passed, is of some concern. These near-term threats are not being pursued, as currently there is no answer with rockets. The basing of Planetary Defense Capability at the Apex Anchor locations will ensure these threats are successfully defeated. The first new concept proposed is a "finding them" issue: Apex Anchor telescopes identifying "near term threats" with stereoscopic vision from Apex Anchors 200,000 km apart staring at the Sun. [see Figure 4.2.2] The second remarkable capability of Space Elevators is that of a planetary defense garage which should be placed at each Apex Anchor to ensure that assembly of stored components, allowing release and rapid velocity towards threats and released within 24 hours.

Apex Anchor Concept: These two elements of the Planetary Defense approach from Apex Anchors would be very successful against near term threats, not seen when coming out of the sun:
- Continuous stereoscopic observations, and
- Assembly and release within 24 hours of threat recognition

Planetary defense systems are crucial in safeguarding humanity's existence on Earth, particularly in the face of potential threats posed by medium to large asteroids. It is necessary to deploy defense mechanisms in a timely manner to protect the planet from potential impacts that could have catastrophic consequences. ISEC introduced the concept of the Apex Anchor Planetary Defense System (AAPDS) by leveraging a Dual Space Access. This research study outlines a comprehensive approach that integrates deep space tracking and Asteroid "Buster" capabilities. [Eddy, 2023] The AAPDS aims to provide a rapid response to near-Earth threats by storing and assembling asteroid busters at several Apex Anchors within a swift timeframe of 24 hours, emphasizing the significance of efficient and effective defense strategies.

---

[2] This Photo by Unknown Author is licensed under CC BY-SA

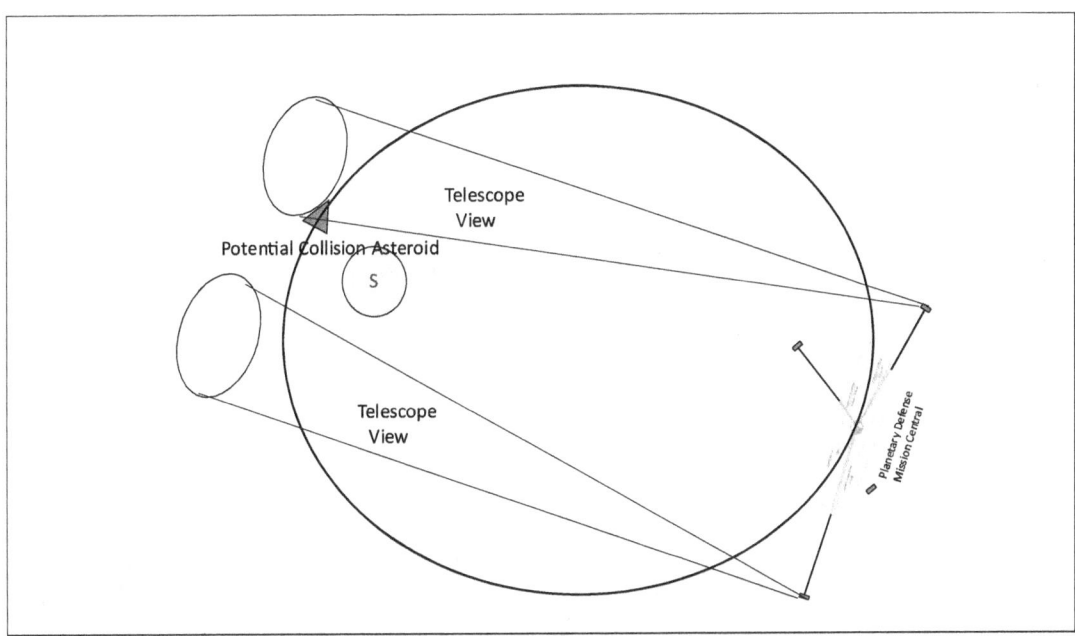

Figure 4.2.2: Stereoscopic Observations for Planetary Defense

4.3 Space Transportation Port (a.k.a "Truck Stop")

Mission: Provide facilities to accept space systems entering the Apex Anchor [from orbits or from the tether], service the vehicles, and release back to orbit or along the tether. In addition, it will release space systems towards destinations after checkout, assembly, repair, recycle, and refueling – as needed.

Figure 4.3.1: Apex Anchor Transportation Lobby

One discussion leveraged the concept of modelling several Earth based logistics centers, such as Naval resupply operations, -- LEO space stations' operational history – and Earth based "outposts" with people passing through, such as Antarctica . In addition, it was brought up during this research that the Apex Anchor would include the reception of incoming space systems that would match the velocity and altitude so they could "just be collected" nearby. This last capability could enhance the concepts of returning valuable products from mining the various asteroids close by or Mars cargo.

In addition, this capability at 100,000 km altitude will also enable release of large space systems with multiple scientific platforms towards far distant destinations for very fast transit. When we talk about scientific missions beyond Earth we always talk kilograms and minimum velocities to planets. [Voyager 1 was 815 kilograms of space craft – at 0.001% of launch vehicle mass - and took 46 years to get where it is today, slowly compared to what can be accomplished with Apex Anchor releases]. Recent estimates of velocity of release from the Apex Anchor can result in at least 29.78 km/sec (as shown in Section 3.4, when combined with Earth's velocity results in relative velocity of 85 Km/s – or enough to leave the solar system with 40 Km/s). In addition, the size of the scientific spacecraft could be in the 1,000-tonne mass region going anywhere within our solar systems, with daily releases. Section 5.2 expands on this concept explaining the potential for a safe harbor and a logistics center.

4.4 Space Construction Center

Figure 4.4.1: Space Construction Platform[3]

Mission: Support and enable construction of facilities and space systems with the ability to build, repair, and improve upon the Apex Anchor so they may be put together with components much less expensive to lift into space than with rockets. It can also be used to build a small space voyager, like the USS Voyager.

Figure 4.4.2: Intrepid Class Spacecraft

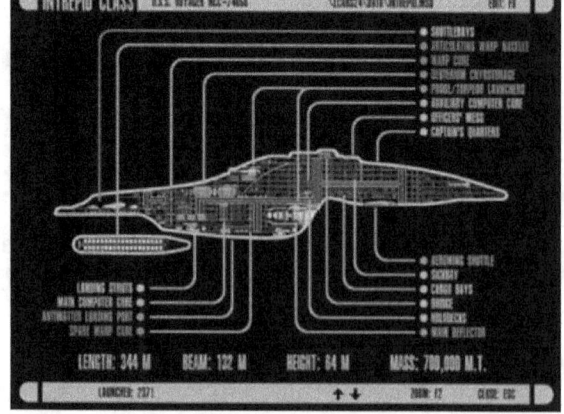

This Photo by Unknown Author is licensed under CC BY-SA

---

[3] This Photo by Unknown Author is licensed under CC BY-NC-ND.

Assembly, storage, and repair of components (segments) of space system so that they may be put together above gravity well. In addition, storage of fuel would enhance the operational flexibility of missions within CISlunar and our solar system.

4.5 Space Hospital

Mission: Attend to the medical needs of the residents and the transient personnel as needed, providing a healing environment alongside science and technology. There could even be a surgical center, although because of the microgravity environment, surgery and wound care in space will be challenging. Some surgery can be conducted remotely.

Figure 4.5.1: Apex Anchor Hospital Ward

This Photo by Unknown Author is licensed under CC

4.5.1 Advantages of a Hospital at the Apex Anchor

Some of the advantages of locating a hospital at the Apex Anchor are:

- Faster evacuation to Earth than a more distant independent station in case of a catastrophic event, i.e., days vs weeks or months
- An environment that will not put additional stress on a body that is acclimated to microgravity [Faini, 2021]
- Ability to get specialized equipment to the hospital quickly on a cargo transport up the space elevator
- Ability to transport specialists to the hospital safely by the Space Elevator
- Medical research, experimentation
- Effective use of gravity for rehabilitation of patients

4.6 Space Logistics, Storage and Distribution Center

Mission: Enable the operations of multiple missions upon the Apex Anchor with storage and distribution of logistics. It is possible to build a large warehouse at GEO, using material raised from Earth on tether climbers since at GEO there is no limit to the amount of material that can be stored there because the region is in microgravity--the forces due to the earth's gravity and the centrifugal force due to rotation about the earth are in balance. The contents of the warehouse will most likely be shipped from Earth also, but it is possible to receive components or raw materials at GEO and build items there. Items shipped to the apex anchor can be quite large, because the centrifugal force due to rotating about the earth is only 5.4% of the gravity here on Earth.

The apex anchor must have enough mass to maintain stability. Because of this, the logistics have to deal with just-in-time transportation of payloads and materials being moved from GEO to the apex anchor. But transportation costs will be minimal due to requiring no significant power.; vehicles naturally accelerate away from Earth beyond GEO.

Figure 4.6.1: Apex Anchor Storage Area

At GEO (the geosynchronous altitude of 35,786km) there is no limit to the amount of material that can be stored there because the region is in microgravity; the forces due to the earth's gravity and the centrifugal force due to rotation about the earth are in balance. Figure 4.6.1 as an example of a logistic center's storage containers. At the apex anchor, the earth's gravity is very much smaller, due to the inverse square law, and the centrifugal force is larger. Thus, there is a net force outward on the tether and any object attached to it. 4.6.2 illustrates the arrangement.

There are many architectural and design options depending on the length and thickness of the tether. In the reference model described in the 2013 book [Swan, 2013], the tether is 100,000km long. To maintain a balance of forces, a mass of 1900 tonnes is needed at the apex anchor. Adding additional mass will increase the load on the tether, and this must be designed into the

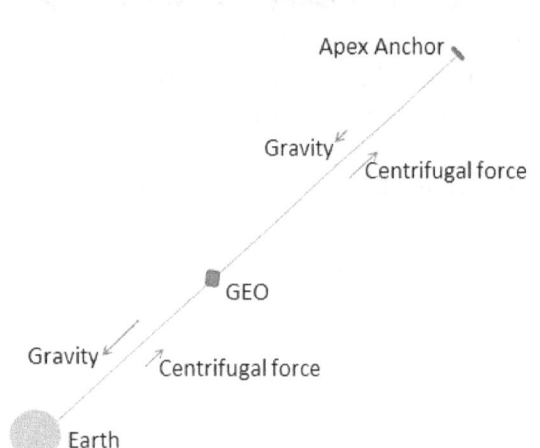

tether by making it thick enough to sustain the extra tension. The extra tension will apply for the whole length of the tether. Removing mass at the apex anchor will reduce the tension, but there is a lower limit beyond which the tether would become unstable. The anchor at the earth's surface must deal with these variations in tension as various loads are added and removed at the apex anchor. This also applies when climbers move up and down the tether.

Figure 4.6.2 Balance of forces along the tether

Once a space elevator has been erected, it is possible to enhance it by adding extra layers of material to increase its capacity for payloads and for facilities at the apex anchor. Indeed, this technique of adding extra layers in situ will be an important step in constructing any space elevator, using the tether itself to add capacity to increase its usefulness, always remembering to maintain the correct ratio – the taper ratio – between the maximum thickness at GEO and the thickness at other altitudes.

The conclusion is that the tether must be designed to sustain the maximum load, but the apex anchor must have enough mass to maintain stability. In some cases, it may be necessary to jettison some material from the apex anchor, from where it will leave the earth's sphere of influence and go into orbit around the sun or will journey towards another planet if correctly aimed.

For these reasons, logistics has to deal with just-in-time transportation of payloads and materials from the unlimited storage capacity at GEO to the apex anchor. Transportation from GEO to the apex anchor requires no significant power; vehicles will naturally accelerate away from Earth beyond GEO. The main requirement is to apply the brakes to limit their speed in such a way that they do not damage the tether and to slow them when they reach their destination at the apex anchor.

For a tether that is 100,000km long, the distance from GEO to the apex anchor is 64,214km. At 382kph this journey will take a week; at 300kph it will take nine days. Therefore just-in-time planning and delivery will be needed to ensure that the correct payloads reach the apex anchor when needed for operations there, such as launching interplanetary spacecraft or building a research station or hotel.

It is possible to build a large warehouse at GEO, using material raised from Earth on tether climbers. The contents of the warehouse will most likely be shipped from Earth also, but it is possible to receive components or raw materials at GEO and build items there, or even build at the apex anchor if required. Items shipped to the apex anchor can be quite large, because the centrifugal force due to rotating about the earth is only 5.6% of the gravity here on Earth (see Table 4.6.1). Indeed, Einstein in his general theory of relativity showed that the acceleration due to gravity is of the same kind as that due to other forces. Thus, the effective weight of a 100-tonne mass is only about 5.3 tonnes. The weight will be experienced as a force away from Earth, and so the earth will appear to be overhead to anyone who is on the apex anchor. Whereas masses lifted from Earth to GEO will be limited by the strong gravity near Earth, masses sent to the apex anchor from GEO could be 10 times greater or even more. This means that a workshop at GEO can assemble substantial structures there for later shipping to the apex anchor.

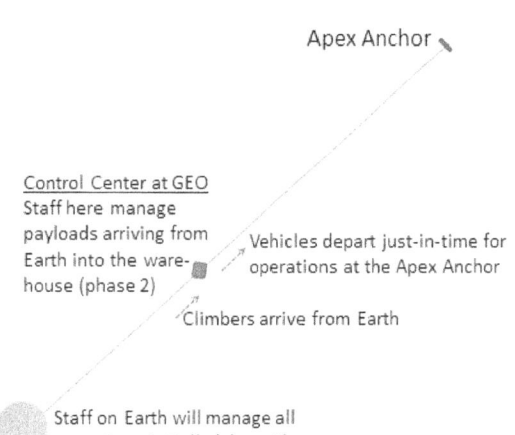

Figure 4.6.3 Management of logistics to supply the apex anchor

It should be possible for logistics to be managed remotely, but the possibility is there for human supervision on site (Figure 4.6.3). There are degrees of automation from complete robotic performance of the tasks through to manual performance. In between, there is remote supervision from Earth or from a control center at GEO or at the apex anchor. However, remote control of activities at the apex anchor from Earth will be subject to a round-trip speed-of-light latency of 0.7 second. Remotely controlling work at GEO from the apex anchor or vice versa will have a latency of 0.4 second. Remotely controlling work at GEO from Earth will have a latency of 0.2 seconds. Latency of this sort is manageable with training and practice. Staff operating remote machinery will be able to learn and practice on simulators. An organist playing a traditional church organ has to deal with similar latencies. We envisage that in phase 1 of the project, all management and operation of equipment will be performed from Earth. Later, when it is deemed safe, phase 2 will commence, and staff will be able to work at GEO or at the apex anchor.

It is unlikely to be economically worthwhile to pressurize an entire warehouse, so direct manual working at GEO or the apex anchor would require spacesuits to be worn. In any case, payloads will have to be loaded and unloaded to and from vehicles in the vacuum. The alternative of working in a safe, comfortable control center and remotely controlling the equipment is probably the best scenario, whether it is close by or at another location thousands of kilometers away. Although the initial operation of the space elevator is expected to be unmanned, just like satellites launched today, we anticipate that people will be able to travel and work at GEO and the apex anchor once the safety and reliability of the climbers and infrastructure have been demonstrated.

4.7 Space Solar Power Distribution Center

Mission: Enable the collection of solar energy and then the ability to distribute around the Apex Anchor and to regional space systems in need of electrical energy. It would encompass building space solar power capable segments that would collect and then radiate energy to tether climbers (from above) and satellites in GEO and beyond, as well as power the Apex Anchor.

They can also be used to Power the Apex Anchor Missions [Woods, 2023]. Could you transmit down the Tether to the Earth Station for Power? Supply energy to space systems within view of Apex Anchor. This would encompass building space solar power capable segments that would collect and then radiate energy to tether climbers (from above) and satellites in GEO and beyond, as well as power the Apex Anchor. (solar cells scavenged from tether climbers?)

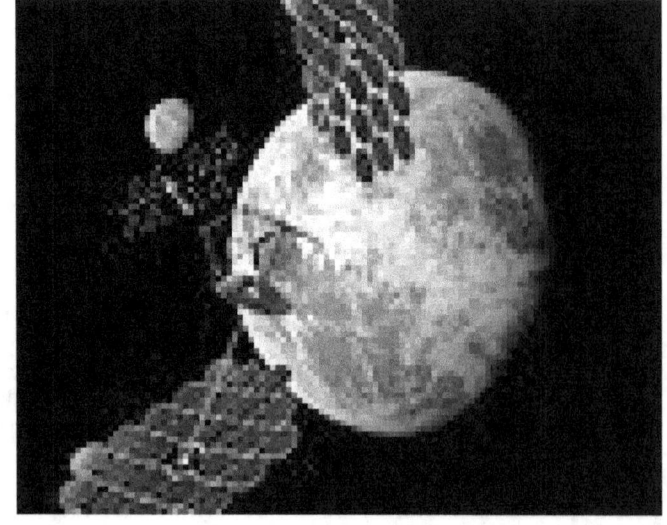

Figure 4.7.1: Space Solar Arrays[4]

---

[4] This Photo by Unknown Author is licensed under CC BY-SA.

International Space Elevator Consortium *ISEC Position Paper #2024*

Figure 4.7.2: Solar power collection and distribution[5]

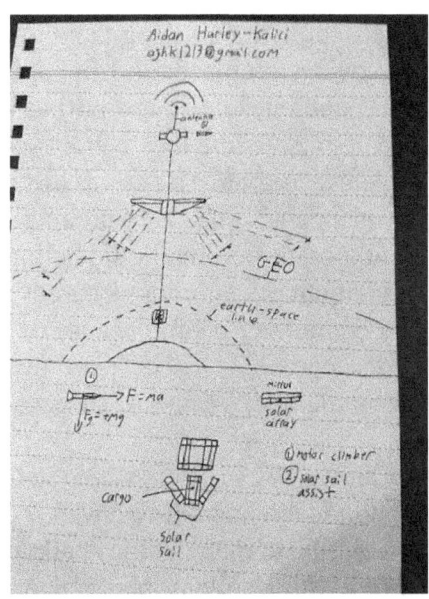

In addition, this capability of radiating photons of light will enable light sail propulsion. With multiple Apex Anchors working together, there could be continuous power propulsion to scientific spacecraft – helping them accelerate towards massive velocity pointed towards destinations such as the next star or just investigations areas between here and there.

4.8 Apex Anchor Astronomical Observatory

Mission: Enable stationary location for observation of the arena around the Apex Anchor, the solar system. and universe.   An astronomical observatory situated far beyond the Earth's geostationary orbit offers a unique solution to many of the challenges faced by current space telescopes. This platform would be equipped with a suite of permanent multi-wavelength telescopes, each optimized for different regions of the EM spectrum to provide a comprehensive view of the universe. From the study of exoplanets and black holes to the exploration of cosmic dawn and galaxy evolution, the platform's capabilities hold the promise of unlocking the secrets of the universe and expanding the frontiers of human knowledge.

 Space telescopes have been significant tools in astronomy over many decades. They provide a platform above Earth's atmosphere which absorbs many parts of the electromagnetic (EM) spectrum: gamma rays, x-rays, ultra-violet (UV), infra-red (IR), and mm to cm radio waves. Without space telescopes the above parts of the EM spectrum could not have been probed and we would have been left with a very limited view of our universe. Space telescopes have literally opened our eyes to gamma-ray bursts, black holes, exoplanets, supernovae, distant galaxies, and the cosmic microwave background radio with all the information encoded in it about the formation of our universe, its evolution, and its basic ingredients. Even the optical part of the EM spectrum greatly benefitted from the incredible Hubble Space Telescope (HST).

Figure 4.8.1: Astronomical Observatory

 Be that as it may, space telescopes come at a hefty price, not only in terms of money. Their

---

[5] Image accomplished in Apex Anchor workshop by Aiden Hurley-Kalici, used by permission.

lifetimes are limited. Servicing them is also problematic. The only serviceable space telescope, the HST, was by design placed in LEO accessible by the then operational US space shuttle fleet. All the other space telescopes are not serviceable. If they malfunction, they cannot be fixed, they cannot be upgraded, and they cannot be replaced unless by another space telescope. For example, IR space telescopes have a natural lifetime linked to the consumption of the liquid helium used to cool its instruments. The hardware might still be in perfect condition, but it would be operationally useless without the cooling liquid helium. The venerable HST had its gyroscopes replaced several times to maintain its pointing accuracy. The powerful Chandra X-ray telescope launched in 1999 is now on the cutting board because of pointing difficulties due to its ailing gyroscopes [Chandra X-ray Telescope, 2024]. These are just two examples of very successful space telescopes being undercut by their very cause of success orbiting high above the atmosphere.

The apex anchor (AA) offers a unique opportunity for a multi-wavelength observing platform. An AA astronomical observatory (AAAO) situated far beyond the Earth's geostationary orbit, offers a unique solution to many of the challenges faced by current space telescopes. This platform would be equipped with a suite of permanent multi-wavelength telescopes, each optimized for different regions of the EM spectrum: IR, visible light, UV, X-rays, and gamma rays.

Unlike conventional space telescopes, which orbit the Earth, the AAAO would be stationed at a fixed position relative to our planet, effectively eliminating orbital constraints. Additionally, its distance from Earth would significantly reduce interference from the planet's atmosphere, enabling unparalleled observations across various wavelengths. The AAAO will operate just like any ground-based observatory where new cameras and equipment could be easily deployed without the need for a full space mission. Moreover, the AAAO could house telescopes of large sizes operating in different parts of the EM spectrum. The current limitation on space telescope size is the current need to have the telescope fit in the launch vehicle. The 2.4m HST was designed that way to fit the US space shuttle's cargo bay. The James Webb Space Telescope (JWST), a 6m IR space telescope located at Earth-Sun L2, utilized an origami unfolding mirror that could fit in the faring of an Arianne 5 rocket. Telescopes part of the AAAO will not suffer from these limitations. They could be built to arbitrary sizes limited only by existing technology. The AAAO telescopes could be upgraded, refurbished, torn down and replaced by more advanced versions, just like ground-based telescopes.

Equipping AAAO with a suite of telescopes covering various wavelengths would provide a comprehensive view of the universe. By combining observations across different parts of the spectrum, scientists could gain deeper insights into astrophysical phenomena, from the formation of stars and galaxies to the behavior of exotic objects such as neutron stars and quasars. The capabilities afforded by AAAO telescopes would open new avenues for scientific discovery across a wide range of astrophysical fields. Here are some examples of the groundbreaking research:

1. Exoplanet Characterization: With its stable observing platform and multi-wavelength instruments, scientists could conduct detailed studies of exoplanet atmospheres, searching for signs of habitability and potential biosignatures. High-resolution spectroscopy in the infrared and ultraviolet regions could reveal the composition and dynamics of exoplanet atmospheres, shedding light on their potential to support life. (e.g., [Kempton, 2024])

2. Cosmic Dawn and Reionization: Observations of the earliest galaxies and quasars are crucial for understanding the cosmic dawn—the epoch when the first stars and galaxies formed—and the subsequent reionization of the universe. AAAO telescopes could capture faint signals from these distant objects with unprecedented sensitivity, helping to unravel the mysteries of cosmic evolution. (e.g. [Klessen, 2023])

3. Black Hole Dynamics: Black holes are among the most enigmatic objects in the universe, exerting gravitational influence on their surroundings and emitting radiation across the electromagnetic spectrum. By combining observations from X-ray, ultraviolet, and visible light telescopes on the platform, scientists could study the accretion disks, jets, and gravitational waves produced by black hole interactions, advancing our understanding of their properties and behavior. Having UV and X-ray telescopes observing the same target simultaneously will shed light on the accretion physics that power active galactic nuclei and help study the symbiotic relationship between supermassive blackholes and their host galaxies. (e.g., [Combes, 2023])

4. Galactic Archaeology: The AAAO's unprecedented observation capabilities would be invaluable for studying the structure and evolution of galaxies over cosmic time. By surveying a wide range of galaxies across different wavelengths, astronomers could trace their formation histories, identify key processes such as mergers and star formation, and map the distribution of dark matter within galaxy clusters. (e.g., [Gauri, 2024])

5. Transient Events: The AAAO's ability to monitor the sky continuously, with the standard architecture with three Galactic Harbors, would make it an ideal tool for studying transient phenomena such as supernovae, gamma-ray bursts, and tidal disruption events. Rapid follow-up observations across multiple wavelengths could provide insights into the underlying mechanisms driving these explosive events, offering clues to the physics of extreme astrophysical processes. (e.g., [Bozzo, 2024])

6. Planetary Defense: This mission will be revolutionary in notification of Earth threatening asteroids by having a 200,000 km observation baseline (two Apex Anchors on opposite sides of Earth) to provide stereoscopic observations. This leads to early identification of Earth threatening asteroids which will "come out of the sun." These telescopes will be multiple wavelengths for broad capabilities staring at the Sun.

The AAAO for permanent multi-wavelength space telescopes represents a paradigm shift in astrophysical research. By overcoming the limitations of Earth-based observatories and conventional space telescopes, such a platform would enable groundbreaking discoveries across a wide range of scientific disciplines. From the study of exoplanets and black holes to the exploration of cosmic dawn and galaxy evolution, the platform's capabilities hold the promise of unlocking the secrets of the universe and expanding the frontiers of human knowledge. As we continue to push the boundaries of space exploration, investing in ambitious projects like this could revolutionize our understanding of the cosmos and inspire future generations of scientists and explorers.

4.9 Hi-Level Nuclear Waste Disposal System

Mission: Allow nuclear waste to be transported from the Earth to the Apex Anchor on its way to a destination (orbit) far away from the Earth for safely disposing of high-level nuclear waste into a solar orbit. At a delivery rate of 170,000 tons/year, the space elevator could safely dispose of all nuclear waste in less than a decade. The feasibility and safety of disposing of the nuclear waste by putting it on a Sun-bound orbit, i.e., throw our nuclear waste into the Sun is also being considered, but would require a heat shield that could withstand very high temperatures and pressures/densities.

Nuclear waste management poses a challenge, especially high-level radioactive waste generated from nuclear fission processes. Significant quantities of high-level radioactive waste are produced globally, both from nuclear weapons production and commercial nuclear electric generators. Given the long-term environmental and health risks, the need for safety and permanent disposal of this waste is of utmost importance.

The U.S. has accumulated a substantial amount of high-level radioactive waste from nuclear weapons activities, with a total of 315 Kt produced from 1945 to 2016. The total amount of nuclear waste is estimated to be 810,000 tons by 2050 (Green Road to Space; GRS). Various disposal methods, including deep geological repositories (DGR), are considered to address the disposal of this waste. Challenges are associated with reprocessing spent nuclear fuel and the importance of ensuring long-term isolation of radioactive waste to prevent human exposure. GRS investigated the use of a space elevator to dispose nuclear waste and noted that the US government as early as 1978 investigated space burial of nuclear waste: high orbits, lunar orbits, or solar orbits, however, the safety concerns of a possible sub-orbital failure and the draconian measures of the rocket equation shelved the idea of outer space as a means of nuclear waste disposal.

However, space elevators can be used for safely disposing of high-level nuclear waste into specific orbits around the Sun. The disposal orbit is a much smaller orbit than Earth's orbit and is designed never to approach Earth again. The process involves raising the waste to 100,000 km altitude and releasing it towards the Sun with a velocity of 7.76 km/sec in the opposite direction of Earth's movement around the Sun. At perihelion (closest point to the Sun), a thrust is applied to reduce orbital energy, decreasing the size of the resulting orbit. Refining orbital characteristics with remaining fuel ensures the waste is stored safely in a disposal orbit, away from Earth. At a delivery rate of 170,000 tons/year, the space elevator could safely dispose of all nuclear waste in less than a decade.

Space disposal of nuclear waste could take many forms. It could be high orbits, lunar orbits, or solar orbits. Given the fact that the Earth-Moon system is going to see increased spacecraft traffic in the near future, high and lunar orbits will not be the ideal locations to store nuclear waste. Accidents are prone to happen. Accidents in space will tend to perturb orbits of resident space objects, such a nuclear waste vessel, and this vessel might come crashing toward Earth or the Moon, which is projected to witness increased human activity with the NASA's Artimis Space Program.

For the above reasons, we are going to concentrate here, instead, on solar orbits as a means of nuclear waste disposal. Solar orbits in the context are orbits around the Sun that are inside Earth's orbit. We are also going to discuss the feasibility and safety of disposing of nuclear waste by putting it on a Sun-bound orbit, i.e., throwing our nuclear waste into the Sun. Solar orbits were presented briefly in GRS (see Fig. 4.9.1 below).

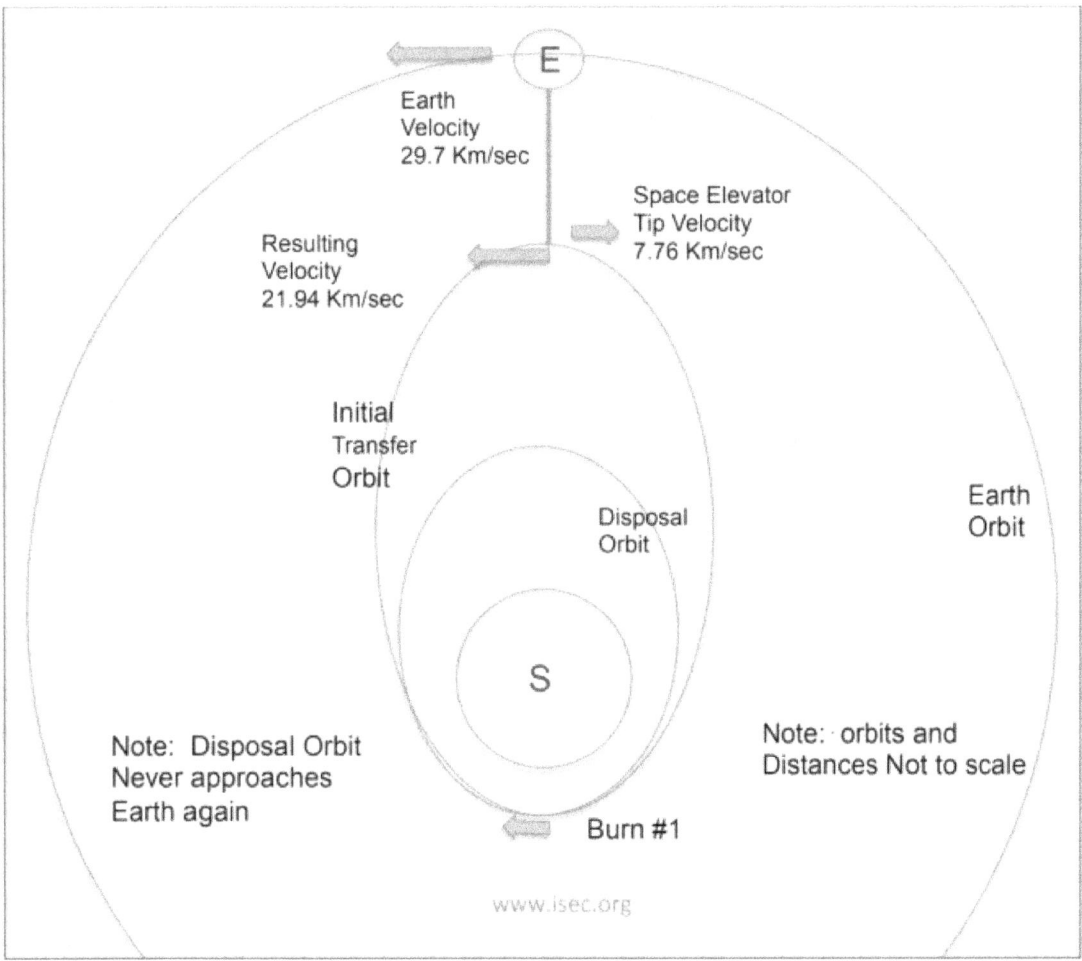

Figure 4.9.1: Sequence for Sun Earth Disposal orbit (GRS).

The nuclear waste vessel climbs the space elevator and is released at a height of 100,000 km in the direction opposite to Earth's rotation. The net velocity, in a Sun centric frame, would be 29.7 km/s less than 7.76 km/s, or 21.94 km/s. This would place the vessel in a transfer orbit. When the vessel reaches perihelion, a burn is initiated to put the vessel on its final disposal orbit.

An added layer of complexity not discussed in GRS is that the initial transfer orbit will not be in ecliptic plane. The space elevator is in Earth's equatorial plane. The release velocity vector will be in the equatorial plane. The initial transfer orbit will be a Sun orbit, but its plane is in Earth's equatorial plane, making 23.5° with the ecliptic plane. Having said this, we emphasize that there is no need to change the orbital plane since we are not targeting a specific solar system object such as a planet or asteroid, which all orbit in the ecliptic plane. Hence, our Sun orbits for nuclear waste disposal will not be in the ecliptic plane. This should be considered as an added safety feature. Orbit-altering accidents are unlikely to be strong enough to change the orbital plane. A nuclear waste vessel is unlikely to be placed in the ecliptic plane through an unplanned perturbation.

Another way to dispose of nuclear waste would be to plunge it into the Sun. However, this venue should be carefully researched since it might result in radioactive waste dust being carried by the solar wind toward Earth. From an astronomical perspective, it is possible to design a Sun-plunging orbit. The Parker Solar Probe is a prime example here. The nuclear waste vessel will in this case have to be equipped with the necessary control thrusters to perform the necessary orbital maneuvers that would put it on a Sun-plunging orbit. With the right sequence of control thruster firings, a Sun-plunging orbit could be achieved after the vessel is placed in the initial transfer orbit.

The safety of plunging nuclear waste into the Sun requires further investigation. Nuclear waste does NOT get "burnt" in the high temperatures of the solar atmosphere. Uranium and Plutonium have vaporization temperatures around 2000-3000°C. A shield vessel, similar to the Parker Solar Probe, will be vaporized well before it reaches the base of the Sun photosphere. This will necessarily disperse the nuclear waste atoms and compounds allowing them to be carried by the solar wind, a dangerous prospect given that the solar system is constantly immersed by the solar wind. Unless the nuclear waste disposal vessel is shielded by a material that can withstand temperatures on the order of 10,000° C, and very high pressures/densities, then plunging nuclear waste into the Sun is not a safe option.

A heat shield would be needed to allow the vessel to penetrate the Sun's photosphere and arrive at the top of the convection zone. The almost million tons of nuclear waste will be diluted deep in enough in the much more massive Sun (2×10^{30}), thus posing no risk further to humanity. The pressure at the bottom of the photosphere is about one million Pascals. For comparison, standard atmospheric pressure on Earth is 100,000 Pascals. The temperature at the bottom of the photosphere is about 5500°. No current material is able to withstand such high pressures and temperatures. The combined effect of high temperature and pressure means that the heat transferred from the photosphere to a heat shield will be immense. No current material is able to withstand these extreme conditions. Kindly note that the Parker Solar Probe has been through the Sun's corona, whose temperature is 1 million degrees. However, the corona's density, and subsequently pressure, is relatively low enough that the specially designed heat shield of the probe is able to survive the corona.

It can be seen from the above, that the safest method of nuclear waste disposal would be to use the space elevator to put nuclear waste in a small Sun orbit.

4.10 Space Hotel at the Apex Anchor [Mafi, 2023]

Mission: Enable transients and residents to have living quarters within the Apex Anchor that are safe and convenient for the work environment. A hotel would obviously support tourism, but the amenities would require a maturation of capabilities and will be planned for fully operational capability estimated to be ten years after initial operations. This would include missions to the Moon and Mars as well as tourism to anywhere in the area; it could attract artists, scientists performing microgravity experiments, people requiring long-term convalescence after major medical events; students and educators, and others seeking to discover the mysteries of space.

This Photo by Unknown Author is licensed under CC BY-SA

Figure 4.10.1 Reception and Room of Space Hotel

When we think about space hotels, what often comes to mind are billionaire tourists hitching rides on rockets to facilities like the International Space Station (ISS) where they get in everyone's way just to have a privileged experience. What if there was a facility made for almost anyone to visit at the apex anchor of a space elevator that was a place of joy and discovery? What if this facility could be a living quarters, a place of entertainment, education, convalescence, events, and recreation? A hotel at the apex anchor could be all of these things and more.

4.11 Next Generation International Space Station

Mission: Empower multiple countries to participate in a space station at the Apex Anchor that would enable a stable environment for government and commercial operations. Observation, communications, navigation, and notification would provide unchartered capabilities from 100,000 km altitude. It might just be the way to bring peace among nations as they work together on major problems or missions that affect all countries or even try to answer the question: Is there life on other planets? We may soon be able to explore and gather evidence beyond our galaxy using space elevators as the green road to space!

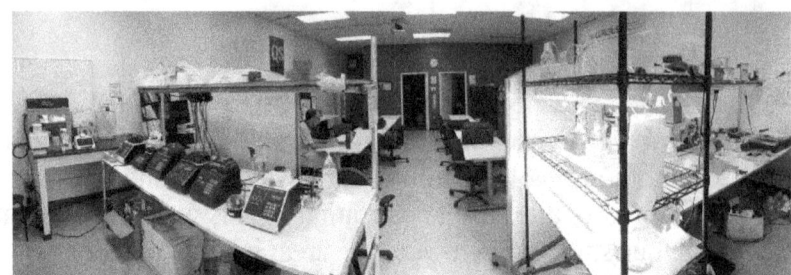

Figure 4.11.1: Labs within the Next Generation International Space Station

High altitude operations will enhance the CISlunar environment as well as help understand the environment from the Earth's surface to well out in the solar system. Observation, communications, navigation, and notification could be functions/capabilities from 100,000 km altitude.

Taking off from Carl Sagan's book, titled "Contact", this might be an opportunity to bring nations into peaceful collaboration by working together on a major project (maybe waste disposal or some other issue that affects the world).

Chapter 5: Apex Anchor Mission Examples

5.0 Introduction

We find ourselves in a new age, in which we are increasingly interconnected, interdependent, and pressed for time. Our ability to predict, and hence to plan, has been greatly diminished as a consequence of the complexity and dynamics of our environments and the nature of the responses necessary to survive and prosper. [Alberts, 2011] For effective mission operations within the complexities of the Apex Anchor, there needs to be a common infrastructure to allow for daily operations. This is essentially a "plug-and-play" type infrastructure.

Figure 5.0.1: Artist Concept of a DSRV[6]

Although the systems can be complex they must all provide standard interfaces so that all systems, no matter the country of origin, can seamlessly integrate. An example is the DSRV (Deep Sea Recovery Vehicle), Figure 5.0.1, that the US Navy created to connect to any submarine, no matter the country, for rescue.

It is envisioned that at some point humans will inhabit the Apex Anchor and conduct missions. Figure 5.0.2 illustrates a conceptual view of an Apex Anchor. This Chapter describes some missions that can be performed within the Apex Anchor.

Figure 5.0.2: Conceptual Illustration of an Apex Anchor, circa 2042.[7]

5.1 Reference Domains

All missions performed within or around the Apex Anchor need to be synchronized for optimal performance. Humans as well as the systems within the Apex Anchor are universal in nature and come from all countries, cultures, and companies on Earth. Over the past 10 years, there have been significant strides in developing cohesiveness within NATO, i.e., getting countries and companies to be

---

[6] This Photo by Unknown Author is licensed under CC BY-SA-NC.
[7] This Photo by Unknown Author is licensed under CC BY-SA-NC

interchangeable. Developing an understanding of how and why things work as they do, or could work, is fundamental to systematically improving functionality aboard the Apex Anchor. [Alberts, 2001] Figure 5.1.1 depicts what is known as a "Bubble Chart"[Alberts, 2001] and how it would relate to the operations within the Apex Anchor.

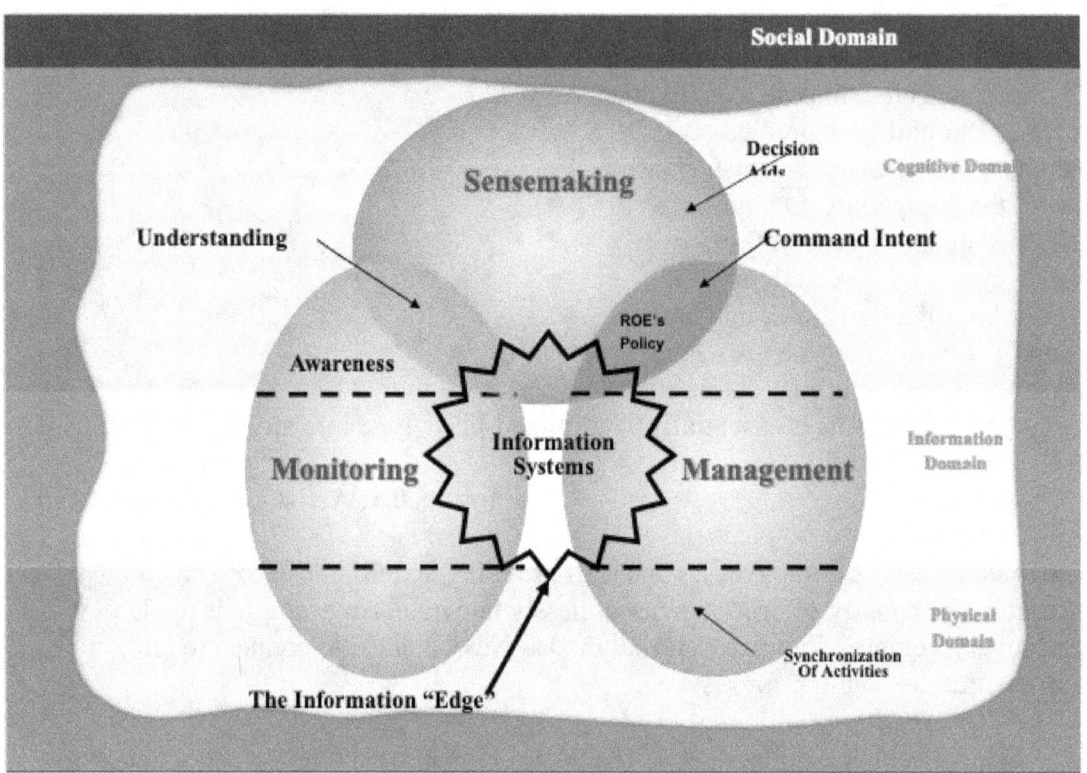

Figure 5.1.1: Multi-Domain Infrastructure [Alberts, 2001]

The interactions among humans and systems can be grouped within the four domains defined as: [Alberts, 2001]
   a. **Physical Domain:** required hardware systems that are robustly networked achieving secure and seamless connectivity and interoperability across different environments.
   b. **Information Domain:** where information has the capability to share, access, and protect information to a degree that it can be created, manipulated and shared among all users.
   c. **Cognitive Domain:** where perceptions, awareness, beliefs, and values reside and where, as a result of sensemaking, decisions are made; and,
   d. **Social Domain:** where individual entities interact, whether verbally or via computer.

Using the four domains, this Chapter describes, in detail, some of the missions that can be performed within the Apex Anchor.

5.2 Detailed Mission Examples

Expanding from Chapter 4, the following provides a more detailed discussion regarding potential missions aboard the Apex Anchor.

5.2.1 Apex Anchor Operations Center

The Space Operations Center is like the bridge on a ship which is tasked to control all operations. Figure 5.1.1 illustrates a conceptual view. It is the control station that is responsible for running all the functions on the Apex Anchor

Figure 5.2.1: Apex Anchor Operations Center[8]

As such, the Operations Center is the heart of all operations within the Apex Anchor. Its primary functions are to: a) develop procedures for the way operations is to be conducted; b) the rules to be employed; c) the relevant organization and doctrine, collaborative arrangements, and information flows; and d) the nature of education and training required.

As shown in Figure 5.2.1, the Operations Center is responsible for three major functions as illustrated in the "Bubble" chart of Figure 5.0.3:
  a. Monitoring: being able to know the status of all systems aboard the Apex Anchor
  b. Management: implementing the "Rules of Engagement" as well as synchronizing all activities aboard the Apex Anchor
  c. Sensemaking: Decision making as well as understanding what is going on throughout the Apex Anchor.

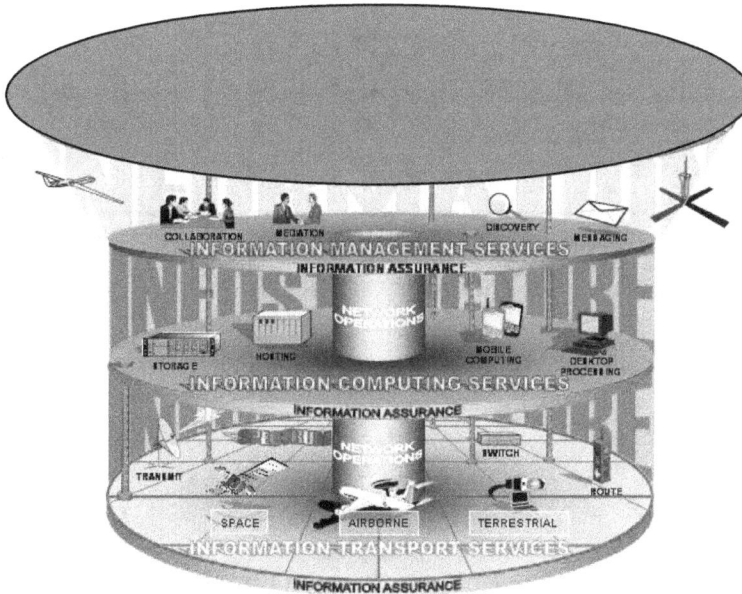

Figure 5.2.2: Network Centric Operations [Phister, 2003]

The Operations Center within the Apex Anchor must integrate all operations within the Apex Anchor. Figure 5.1.2 provides a conceptual view of what is regarded as "Network Centric

---

[8] This Photo by Unknown Author is licensed under CC BY-SA

Operations" to create what could be referred to as the "Infosphere".

The operations within the Apex Anchor are complex. Considering operations within the four domains as discussed in Section 5.0, Figure 5.1.3 provides some of the human interactions necessary to support the Apex Anchor operations.

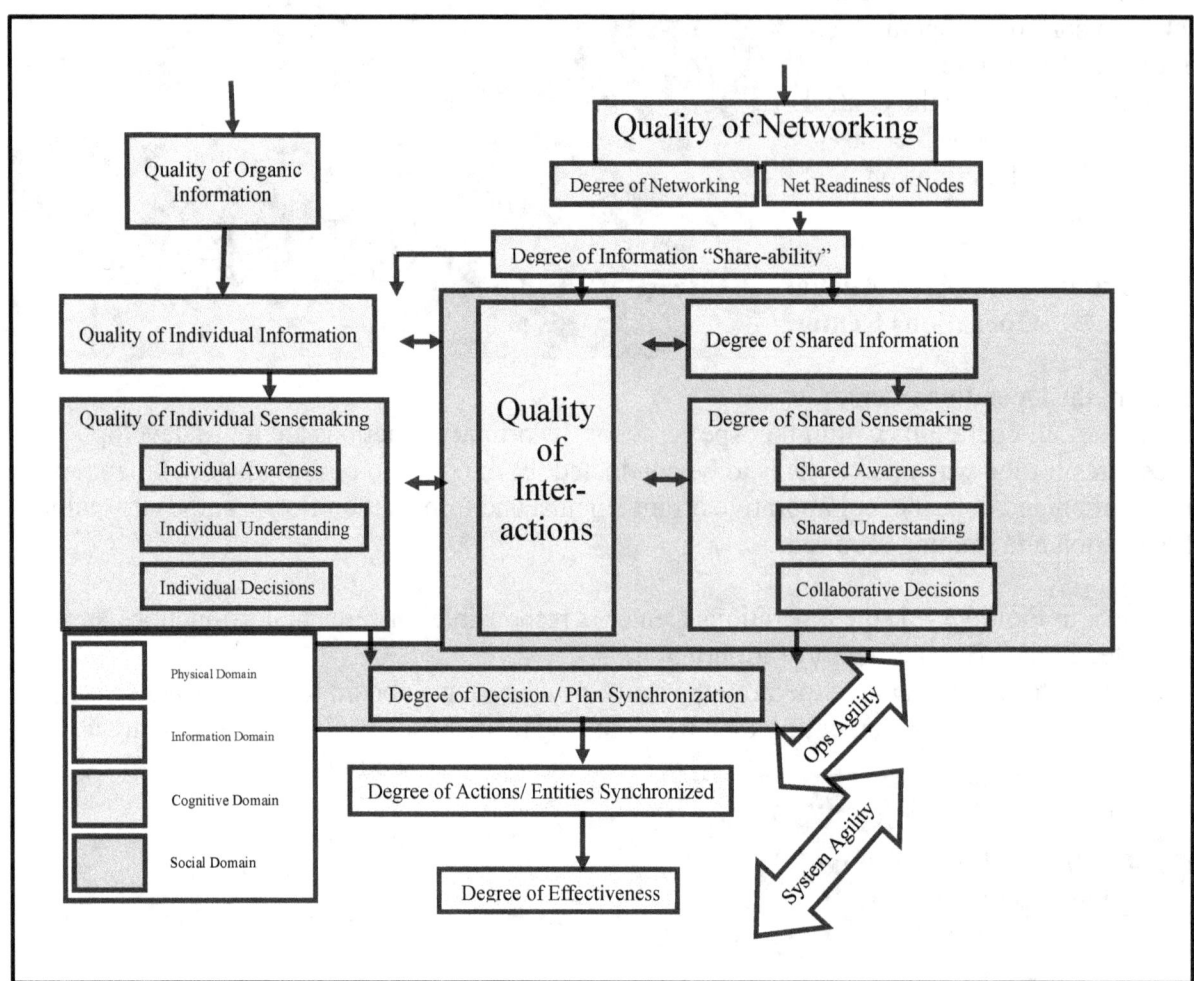

Figure 5.2.3: Netcentric Operations Framework [Alberts, 2003]

Awareness: Cognizant of all aspects within the Apex Anchor, such as: machine status, maintenance status, etc.
Understanding: Individuals within the Operations Center must know their environment in order to understand not only operating procedures but to be able to detect faults before they become an issue.

In summary, the Operations Center is the "heart" of the Apex Anchor and the individuals need to have total "awareness" of the environment and the "Understanding" to deal with all situations.

5.2.2 Intergalactic Transportation System (a.k.a. Galactic Space Train)

> "Now that we have decided to go to the Moon and on to Mars in a combined international, governmental, and commercial effort of great magnitude, we need to expand our vision of 'how to.' It would seem that the establishment of a Full-Service Transportation Node with an ability to assemble, repair, build and store at a location essentially beyond gravity would open up Human movement beyond our planet." [Swan, 2022]

This hypothesis establishes the basics of this discussion. The Apex Anchor will become a Full-Service Transportation Node enabling a location to become – indeed – a "truck stop." The ability to service customers at the top of the Space Elevator will enhance the commercial [and government] missions in a manner that cannot be established with rockets. This safe harbour before release towards distant locations and acceptance of incoming space systems will enhance the mission value for customers. Transportations systems must not only move logistics traffic, but also ensure that the management, repair, refueling, storage, or assembly characteristics of a "truck stop" can be operationally enabled.

When one looks at a transportation system they have wonder about the centuries of improvements that have occurred since the first roads, such as the Roman highways and aqueducts. The definition for transportation infrastructure includes: "It refers to the framework that supports transport system. Transport infrastructure consists of the fixed installations including roads, railways, airways, waterways, canals and pipelines and terminals such as airports, railway stations, bus stations, warehouses, trucking terminals." [web, 19Jun24] When one talks about the Apex Anchor at the top of a 100,000 km Green Road to Space, the excitement starts to rise in customer designs and needs. If one looks at "truck stops" as safe havens in the harsh world of space with the ability pause, refuel, repair, assemble, or just transfer from orbital paths to a highway to/from Earth. These capabilities are not available with rockets with compatible equivalent delta velocities or altitude above gravity. This unique location with its special characteristics will enable a routine methodology to occur daily for movement of massive payloads to anywhere in the solar system with timely releases resulting in massive speeds to shortcut the trips to their destinations.

Provides a "Truck Stop" to the nearby planets and eventually to other parts of the Solar System. For mankind to truly explore the stars, we must develop the capability to lift massive amounts of materials into space. This will require a solution which enables us to deliver these materials much more efficiently than the current vehicles available, such as SpaceX's Starship. Imagine what would be needed to be able to deliver a million people to Mars. The trip from Earth to Mars can take 61-400 days, depending on orbital relationships at the time. Picture being confined to SpaceX's Starship for that period! In addition to all the physical resources that must be handled, passengers would need a room to be able to live; to eat, sleep and work in a comfortable environment during the trip. With the concept of a Galactic Harbour comes the recognition that movement off-planet will require complementary capabilities, rocket portals and Galactic Harbour infrastructures, each with their own strengths and shortfalls. This paper discusses the Galactic Harbour concept as part of the Dual Access to Space Architecture and provides an initial approach to have a continuously moving set of connected vehicles, called an inter-planetary transportation system, or Galactic Space Train, to provide a transportation infrastructure to ferry people, materials and supplies between the Earth and Mars.

5.2.2.1 Strengths of Apex Anchor

One characteristic that will dominate over other transportation nodes is the ability to release at high velocity after being serviced at the truck stop. The next figure shows the various velocities a customer can choose to go to their chosen destination. The standard design is 100,000 km altitude resulting in 7.76 km/sec [or 17358.63 mph] release towards their destination resulting in 14 hours to lunar orbit or as fast as 61 days to Mars. When one climbs further, the velocities increase rapidly as shown.

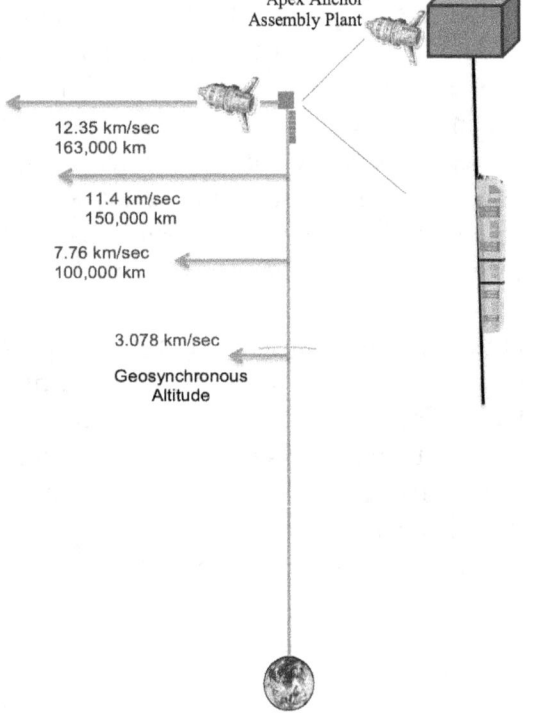

Figure 5.2.2.1.1: Velocity vs Altitude

When leveraging a Full-Service transportation infrastructure, the characteristics jump out to emphasis the amazing strengths. They are:
- Daily, routine, safe, inexpensive and efficient
- Full-service with assembly at the top of the gravity well with refuel, repair and storage
- Unmatched delivery statistics to the Apex Anchor [70% of pad mass]
- Unmatched massive logistics movement
- Green Road to Space
- Unmatched high velocity at release [Delta V]
- Transforming economics towards strategic investments
- Smooth ride – no rocket shake/rattle/roll

This transportation infrastructure will alter the methodology of going to space by paralleling the historic development of roads, cannels, and railroads. Indeed, this Truck stop will enable so many missions for the future that it will be called the premier Space Port.

5.2.2.2: An Example: Radical Mission Option Improvement

One of our greatest achievements was to put the Perseverance Mars lander (with helicopter) on the surface of Mars. The dry mass of the space system was **3,649** kilograms (**8,045** lb.). The delivery statistic was less than one percent of pad mass made it to the surface of Mars. When one uses the Apex Anchor as the release point towards Mars several major changes occur driven by the above characteristics.
1. The velocity at release is huge with the ability to reach Mars - as fast as 61 days with the rest of the year varying between 400 to 70 days. However, the critical part is that with

these velocity choices you may release towards Mars any day of the year – resulting in different trajectories that will not be optimized for minimum velocities as that is not necessary – thus releasing every day enables a transportation characteristic of 'on time' delivery, even have a train schedule for planning.
2. The mass to be released as a scientific spacecraft going to a planet (Mars) or an asteroid can be huge. In the history of humanities missions beyond Earth, the delivery statistic has been less than one percent of pad mass as the mission needed velocity and they had to escape the gravity of Earth by burning rocket fuel. When a large space system is released from the Apex Anchor transportation node, huge missions can be achieved as the mass beats gravity by climbing with electricity to the Apex Anchor.
3. Assembly at the Apex Anchor is a transportation characteristic not allowed by rockets as they have to climb against gravity by consuming fuel and rocket parts. When one has a Full-Service Transportation node, the variations of mission, orbital direction released into, velocities chosen, and day of release are chosen by the customer to enhance their mission. Can you imagine building the starship Voyager at the Apex Anchor and releasing towards any part in our solar system with high velocity. This is a unique strength that rockets cannot match.

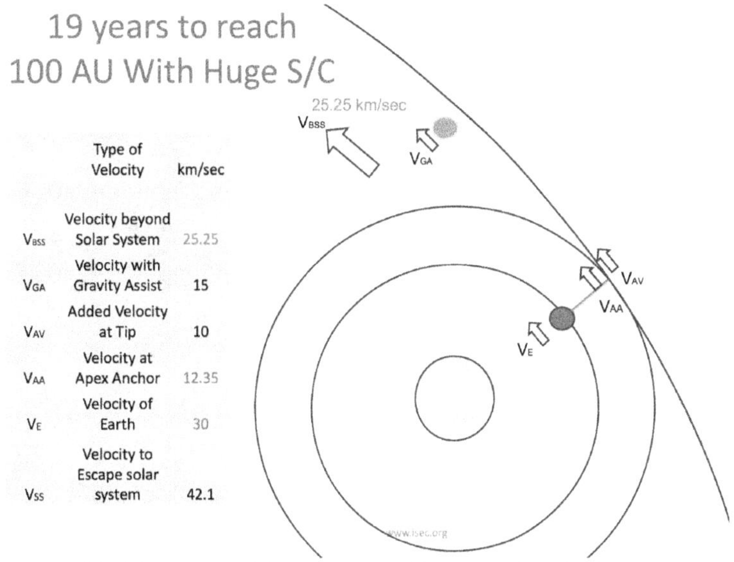

Figure 5.2.3.1: Escape Solar System

Type of Velocity		km/sec
V_{BSS}	Velocity beyond Solar System	25.25
V_{GA}	Velocity with Gravity Assist	15
V_{AV}	Added Velocity at Tip	10
V_{AA}	Velocity at Apex Anchor	12.35
V_E	Velocity of Earth	30
V_{SS}	Velocity to Escape solar system	42.1

19 years to reach 100 AU With Huge S/C

5.2.2.3 Interstellar

The conclusion is that the future interstellar (solar system) scientist may have any size space system assembled before release to beyond the solar system with great velocity. To illustrate this concept, Figure 5.2.3.1 shows some numbers related to a fast release from the Apex Anchor, and with help along the way with gravity assists, can reach interstellar space in a decade. The explanation is as follows: Vbss is the velocity once beyond the solar system; Vga is velocity gained by gravity assist with planets along the path; Vav is the added velocities from the various end effects at the Apex Anchor [see chapter 3]; velocity at tip of Apex Anchor (163,000 km altitude); velocity of Earth; and Vss velocity to escape the solar system. Indeed, the scientist's space system can escape the Solar System gravity with sufficient velocity to reach well beyond our current reach.

5.2.2.4 Mars Cycler Support

Two of the authors of this research report presented a paper in Room Journal with the concept that space elevators fit within Dr. Aldrin's Cycler design. [Phister, 2021] His concept was one where multiple large cruise ships would be placed in solar orbits that periodically approached both Mars and Earth in a single orbit. While close by, the shuttles from the Cyclers would shuttle people and supplies between the planet and the orbiting cruise ship. The concern over the last 40 years has always been that the velocity needed near Earth was very large and could not easily be reached with a "taxi" that was to return to the original planet and beat gravity along the way with rocket motors. When one does the approach numbers from the Cyclers as they approach the planets, the conclusion comes that the velocities match the rotational velocity of the space elevators on Earth and Phobos. As the velocities match for a short period of time, taxis leave the Cycler and the tether – pass in the open space – then rendezvous with the other transportation methodology. The taxis stay with the huge Cyclers and are used at the next space elevator and Cycler rendezvous. This matching of velocities by the Space Elevators and the incoming Cyclers enables the Space Elevator transportation system to act as a Truck Stop transferring cargo and people in a logistically simple manner. The image was used in the Room Journal and reflects this transfer from tether to Cycler and return at the next planet.

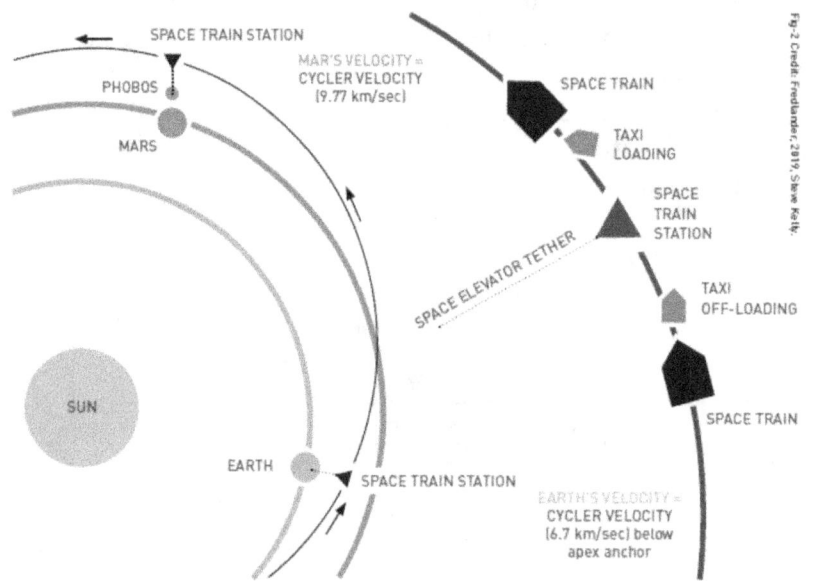

The proposed Space Train would be in a continuous 'Cycler Orbit' (left), as described by Buzz Aldrin and would operate like a continuous train between Earth and Mars (right), the transfers to and from the train being accomplished using velocity-matching Taxis

Figure 5.2.2.4.1:   Mars Cycler Support[9]

5.2.2.5 Going to Mars Utilizing the Inter-Planetary Transportation System

The "Inter-Planetary Transportation System", or Space Train, has three major sections as shown in Figure 5.2.2.5.1:

1) Space Elevator to/from the Earth's Surface to the Apex Anchor Train Station. This can occur daily. The taxi with cargo or people raises itself to the Apex Anchor, is released and then matches speed with the Space Train, attaches itself and supports the trip towards Mars.
2) Utilizing a Space Train to transit to/from the Earth and Mars. This is a continuous cycling system where the Space Train would continually cycle from Earth and Mars. Due to orbital speeds, arrivals and departures would be accomplished using taxis.
3) Then the Phobos Space Elevator catches the taxi released from the Space Train and connects the Mars Train Station to/from the Mars surface.

---

[9] Credit to Fred Lander, 2019 Shawn Kelly Room Magazine [Phister, 2021]

This concept provides a "Green Way" to space utilization [Swan 2020]. The Earth Port would act as the initial embarkation point to load/unload people, supplies and materials for the trip to Mars.

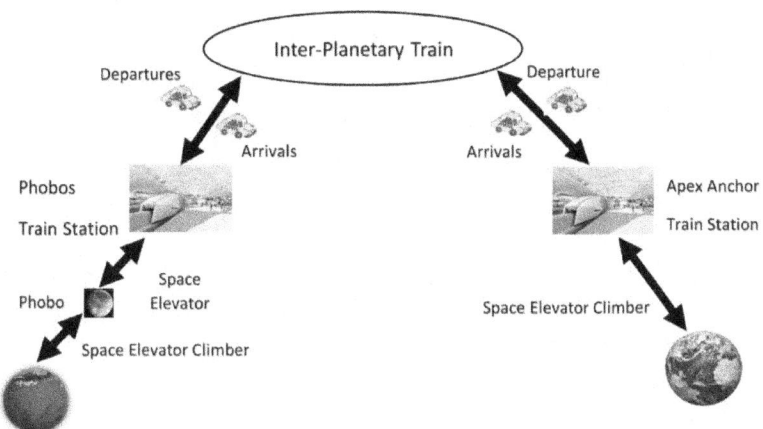

Figure 5.2.2.5.1: Inter-Planetary Train Route

For interplanetary space travel, the Space Train, which is like the trains on Earth, needs to have facilities for the passengers to eat, sleep, relax, and work areas if desired (See Figure 5.2.2.5.2). Why? Because the trip to Mars is, on average, 115-days. One cannot expect passengers to sit (as when on a bus or a rocket capsule) for this length of time. They must be able to sleep, eat, and relax (as when on a train) during their trip to/from Mars. [Musk 2017] The overall concept of the Space Train can be seen in Figure 5.2.2.5.3.

Dining Area

Work Area

Lounge Area

Sleeping Area

Figure 5.2.2.5.2: Living Areas within a Space Train[10]

---

[10] These Photos by Unknown Authors is licensed under CC BY-NC-ND

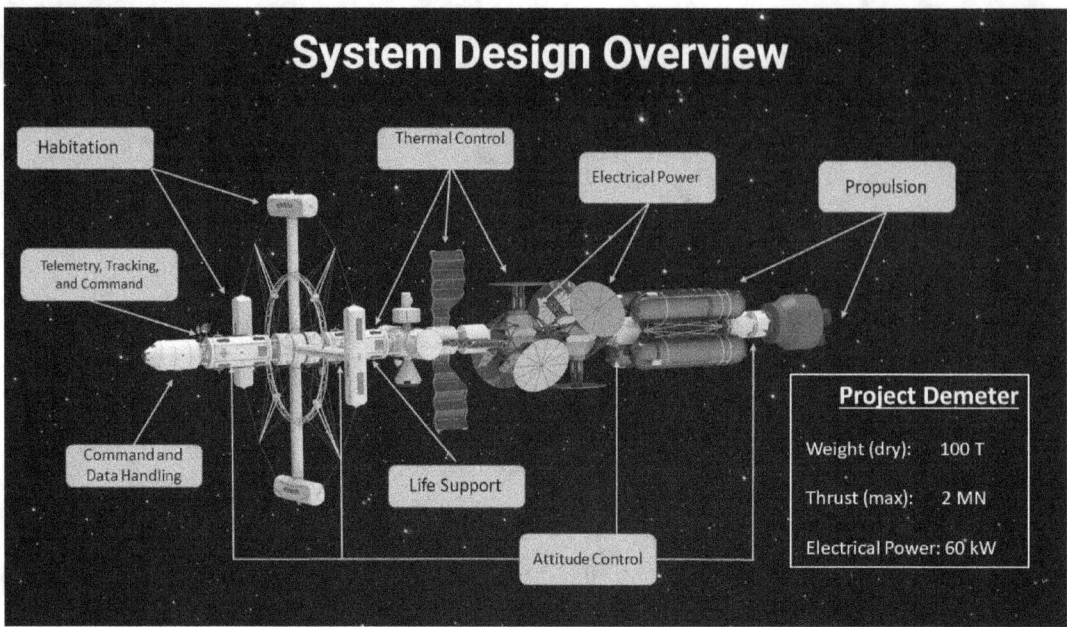

Figure 5.2.2.5.3: Overall Space Train Concept[11]

5.2.3 Planetary Defense

Within the realm of planetary defense, the Planetary Defense Coordination Office (PDCO) within the National Aeronautics and Space Administration (NASA) emerges as a major actor. This office plays a pivotal role in managing various activities related to planetary defense, focusing on essential functions such as searching, detecting, tracking, characterizing, planning, and coordinating responses to objects approaching Earth. By emphasizing proactive measures to identify and mitigate potential threats, especially from asteroids that could pose a significant risk to the planet, the Dual Space Access Strategy [Eddy, 2003] recognizes the importance of early detection and swift action by Apex Anchor planetary defense systems in ensuring planetary safety. Moreover, this ISEC research study sheds light on the expansion of threat arenas, highlighting scenarios where asteroids pass Earth undetected or alter their orbit near the asteroid belt, underscoring the need for enhanced surveillance and response capabilities to address evolving threats effectively and rapidly.

In exploring innovative solutions to bolster planetary defense readiness, the Dual Space Access Strategy ISEC Study [Eddy, 2023] delves into the potential of leveraging Apex Anchor capabilities in planetary defense. By harnessing these cutting-edge rapid response systems, the goal is to streamline the process of deploying defense mechanisms and enhancing overall preparedness to respond swiftly and accurately to asteroid threats. The AAPDS control center has the capacity to automate responses to asteroid threats and ensure seamless coordination of world-wide defense measures. The system's ability to pre-position and configure asteroid busters for diverse threat scenarios is a key advantage over conventional methods, enhancing the system's overall efficacy in mitigating potential risks and safeguarding the planet.

---

[11] Space Train with gravity for astronauts with nuclear power and propulsion (provided by students at Arizona State University in their Capstone Engineering Design Course – Cody Halminiak, Dylan Leigh, Kimberly Luna, Ryan Owen, Kristian Peterson, and Christopher Washburn).

AAPDS fortifies Earth against asteroid impacts and other celestial hazards. It enables a comprehensive approach to planetary defense that prioritizes early detection, rapid response capabilities, and the integration of advanced technologies to bolster defense readiness. The AAPDS is a robust and comprehensive defense system to protect the planet and ensure the continued survival of humanity in the face of cosmic challenges. Of course, the AAPDS is going to deploy a set of telescopes equipped with sensitive cameras to scour the sky for rogue asteroids. These telescopes will necessarily be in the Apex Anchor as part of an astronomical observatory (see Section 4.8). The vantage point provided by the Apex Anchor will deliver stereoscopic capabilities staring at the Sun that will rival all asteroid-hunting, ground-based and orbit-based, telescopes.

The AAPDS will need to launch the asteroid busters in the ecliptic plane. The ecliptic plane is the plane in which Earth, the planets, and all asteroids orbit the Sun. The space elevator and the Apex Anchor are in Earth's equatorial plane. The equatorial plane makes an angle of 23.5° with the ecliptic. The velocity vector of the Buster at release from the Apex Anchor is 7.76 km/s within the equatorial plane. The Asteroid Buster must either have its own control thrusters to effect an orbital change or it must be released from a pivoting ramp attached to the space elevator. (see Chapter 3) The pivoting ramp must be oriented in such a way to insert the Asteroid Buster in the ecliptic plane. This will be the Tier 3 ramp discussed by Peet [Peet, 2021], and discussed in Chapter 3.

Over and above the need to have the Asteroid Buster inserted in the ecliptic plane, the Asteroid Buster must have its own control thrusters to maneuver to intercept the rogue asteroid. The interception must result in a clean deflection in the sense that the rogue asteroid must be deflected away from the Earth. The AAPDS will have to necessarily include a garage at the Apex Anchor to store the booster thruster, control thrusters, and payloads of various hardness and masses.

This design approach is essential to deal with rogue asteroids on a case-by-case basis. This approach would not be possible without the AAPDS. The AAPDS is not just about hurtling rockets at asteroids. The AAPDS is an integrated system of telescopes and sensors to detect and characterize orbits, and astrodynamics package that will calculate interception trajectories and provide flight profile for various options. Each unique mission will have an optimal trajectory and selected payload to affect a clean deflection. A robotic assembly plant at the Apex Anchor will build the desired Asteroid Buster according to the required delivery design.

The AAPDS elements will mostly be located at the Apex Anchor with asteroid-hunting telescopes and the robotic assembly plant. An astrodynamics package would be used to calculate the needed payload for the Asteroid Buster. This mission planning is not so far-fetched if we remind ourselves that the assembly plant is robotic. Needless to say, the Apex Anchor Operations Center will be able to perform AAPDS operations or, if necessary, can be overseen by human operators conveniently located at various centers around the globe.

As mentioned in Chapter 4, Humanity's existence on the Earth depends upon Planetary Defense systems and their timely deployment. The authors believe the fact that several asteroids pass the Earth from the Sun side without being identified until after they have departed is of some concern. The basing of Planetary Defense Capability at the Apex Anchor locations will ensure these threats are successfully defeated. The first new concept proposed is a "finding them" issue: Apex Anchor telescopes identifying "near term threats" with stereoscopic vision from Apex Anchors 200,000 km apart staring at the Sun. The second remarkable capability of Space Elevators is that of a planetary defense garage should be placed at each Apex Anchor to ensure that assembly of stored components could be accomplished in a rapid manner and released within 24 hours towards the threats.

> *Apex Anchor Concept:* These two elements of the Planetary Defense approach from Apex Anchors would be very successful against near term threats, not seen when coming out of the sun:
> - Continuous stereoscopic observations, and
> - Assembly and release within 24 hours of threat recognition

The Apex Anchor Planetary Defense System (AAPDS) will need to launch the asteroid busters into the ecliptic plane. However, the space elevator, and the Apex Anchor, are in Earth's equatorial plane. These two planes only coincide twice a year at the vernal and autumnal equinoxes. Of course, the AAPDS cannot afford to wait till the upcoming equinox to release the asteroid buster. Indeed, the asteroid in question is probably on a trajectory that will make it hit Earth well before the upcoming equinox. Obviously, the AAPDS must have the capability to launch in any plane and direction. This capability is essential to meet any threatening asteroid. In some situations, the safest option might be obliterating the offending asteroid by buster hitting in a such a way so that the resulting debris is directed out of the plane of the solar system. The AAPDS will necessarily have a built-in protocol that advises on which method to use given the Time-for-Arrival (TFA). It is important to keep in mind that almost all studies on planetary defense currently assume the need of at least a year or two of preparation for a mission launched from Earth. The leverage that Apex Anchors provides is that it is platform to launch missions on a moment's notice, provided that storage and assembly facilities are already in place at the AAPDS.

For the sake of definiteness and brevity and since the present study is not an exhaustive exploration of asteroid threats, we are going to adopt the definition of a "threatening asteroid:" A threatening asteroid is an asteroid whose diameter is in the range of several hundred meters. Asteroids with diameters less than 100m will not lead to extensive damage, while kilometer class diameters have their orbits well-mapped out and are also unlikely to suffer sudden perturbations that might put them on a collision course with Earth. The concept of "asteroids out of the Sun" does not fit inside NASA' current strategy. Figure 5.2.3.1 shows NASA definition of hazardous near-Earth objects, as presented in the NASA Planetary Defense Strategy and Action Plan released in 2023.

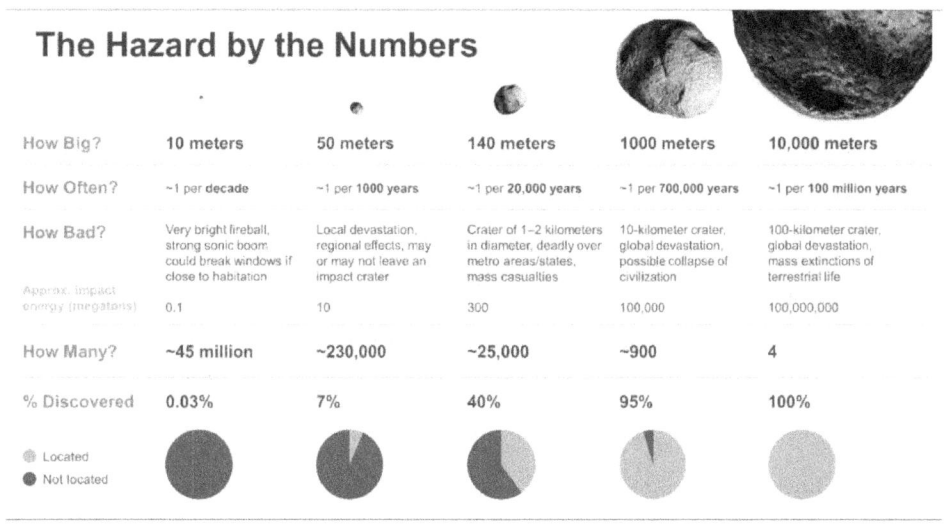

Figure 3: *NEO size and hazard. (Credit: Johns Hopkins University / Applied Physics Lab)*

Figure 5.2.3.1 NASA Planetary Defence Strategy and Action Plan (2023).

Three mission concepts will be discussed in this study. The description of the scenarios is at the top level and back-of-the-envelope calculations level. Going into the deeper technical details will require full-fledged flight mechanics simulations. These missions aim to mitigate the threat of an incoming asteroid by leveraging the capabilities of the AAPDS. Each of the mission concepts will deal with a different scenario dictated by a different asteroid.

- The first scenario is an asteroid such that its perturbed orbit would make it collide with Earth in matter of a couple of weeks.
- The second scenario is an asteroid such that its perturbed orbit would make it collide with Earth in matter of several months.
- The third scenario is an asteroid such that its perturbed orbit would make it collide with Earth in a matter of a several years.

The first detection of the asteroid is already when it's threatening Earth with the time to collision given above. The common strength within AAPDS focused scenario is its ability to respond in real time to near term threats requiring almost instant response as soon as the threatening asteroid is detected. This constraint is easily met by the AAPDS (see Chapter 4.2 in this study), making the almost on-the-fly response to the challenge of deflecting an asteroid a unique AAPDS capability. This study assumes that the AAPDS can deploy asteroids busters in any direction. This could be achieved using secondary tethers, described in Chapter 3 [Knapmann 2021] [Swan 2021] and/or skyhooks and slingshots [Peet 2021] . Regardless of the method used, we assume here the asteroid buster will have a speed of 10 km/s upon release from the Apex Anchor. The various release methods each have a different release velocity, but all are in the ballpark of 10 km/s. We also assume that the asteroid buster, regardless of the technique used to mitigate the asteroid threat, will have its own control thrusters to help it adjust its trajectory to the threatening asteroid.

Considering the first scenario: Detecting an asteroid on a collision course with Earth within a couple of weeks. The AAPDS sensors and detectors in cooperation with the Apex Anchor Astronomical Observatory (AAAO) (see Chapter 4.8) detect a threatening asteroid. The time of flight is too short for any gentle, perturbation measures to be effective. The AAPDS, using its built-in protocol, almost instantaneously determines that this asteroid can be obliterated using a projectile type of weapon (e.g., iron pellets, rubble pile, solid rock, etc.).) (referred to as the payload). The AAPDS will have in stock delivery vehicles that will carry the projectiles to their destination. A supercomputer will calculate the optimal trajectory for the asteroid buster all the way from release at AAPDS to the threatening asteroid. A separate supercomputer on Earth, helped by a pre-defined protocol, will calculate the payload required to pulverize the asteroid. The payload type will depend on the asteroid nature. The calculations must be finalized within the first twelve hours after detection. The next twelve hours will be used to assemble the package and then couple the delivery package to the "Asteroid Buster" located on the Apex Anchor. The plan is to impact the first package within the first week and then have the second package on standby at the AAPDS in case the first package fails to lead to the desired effect.

The asteroid buster will be maneuvered very early after its release to be placed on a shortest-time, head-on collision course with the asteroid. For example, if both the asteroid buster and the asteroid have speeds of 10 km/s and are on a head-on course, then the package will be delivered in one week. Asteroid buster will meet up with the threatening asteroid at distance of 0.04 Astronomical Units (A.U.) from Earth, which roughly 20 times the average Earth-Moon distance.

Depending on the actual size and composition of the asteroid, the "Asteroid Buster" could use multiple approaches to divert or destroy it. The ejecta will serve as a rocket eject and deflect the asteroid away from Earth. The collision in this case should take place sideways, i.e., the asteroid buster should maneuver to fly past the asteroid and intersect just at the right moment – a head-on collision course is still desirable to ensure fast delivery. Alternatively, the asteroid buster can hit the asteroid head-on, excavate a crater because of its kinetic energy and cause destruction. This result will be pulverization of the asteroid to smaller pieces, especially if it is of the "rubble-pile" type.

"Asteroid Busters" could be reconfigured within the "garage" of the Apex Anchor. The payloads would be such that they are large enough to either obliterate or deflect threatening asteroids. Depending on the asteroid, the buster can actually have several smaller payloads that hit the asteroid in well-timed explosions to produce the desired effect. Smaller, payloads are more flexible than the very large ones. Moreover, the form factor and mass of small and medium-size payloads makes them ideal to fit in the delivery spacecraft. Indeed, many of these payloads were designed to fit in rockets (e.g., DART) or to be delivered by planes. The high-energy payloads could be used with the rare, larger asteroids.

Considering the second scenario: Detecting an asteroid on a collision course with Earth within several months. Again, as in the previous scenario, we assume that AAPDS can release asteroid busters in any direction with a speed of 10 km/s, that deciding on what the mitigation method will occur almost immediately according to a pre-defined protocol, and that release of the asteroid buster will be within 24 hours from detection. In principle, one could use smaller payloads as in the first scenario to neutralize this threat, however, larger payloads are relatively expensive and should be spared to deal with clear, present, and growing large threats. The several months' timeframe is still too short for the gentle gravitation methods to be effective. A time of flight of several months is the trickiest timeframe. A time of flight of several months, i.e., eight, and if the asteroid is incoming at a speed of 10 km/s then the threatening asteroid was at distance of 0.65 A.U. at the moment of detection. The AAPDS needs to deliver the payload package within two months of detection, to allow enough time for the deflection method to work. It's assumed that payloads are ready for impending danger within a couple of weeks' time. Very large payloads are basically reserved as a last resort. An asteroid buster going at just the release velocity of 10 km/s will take four months to arrive at the threatening asteroid, assuming a head-on course. There would not be enough time to effectively deflect the asteroid without the use of large payloads.

Considering the third scenario in which the time of flight is several years. Given the long flight time, a variety of methods could be employed by leveraging the AAPDS capabilities. The AAPDS platform located in the Apex Anchor could house a powerful laser to ablate the surface of the threatening asteroid. The evaporated material will act like rocket exhaust with very low thrust but very high specific flux, assuming that the laser operates for a very long time on the asteroid [Lubin, 2015] any of the technical challenges of operating a powerful laser system on a satellite will be circumvented by placing the laser at AA. Figure 5.2.3.2 shows the laser on time need to deflect by two Earth radii an asteroid of given diameter. It can be seen that a 500-meter diameter asteroid can be deflected on a timescale of 3 years by a 1 MW laser. The size of a laser system of this power can easily be assembled at the Apex Anchor.

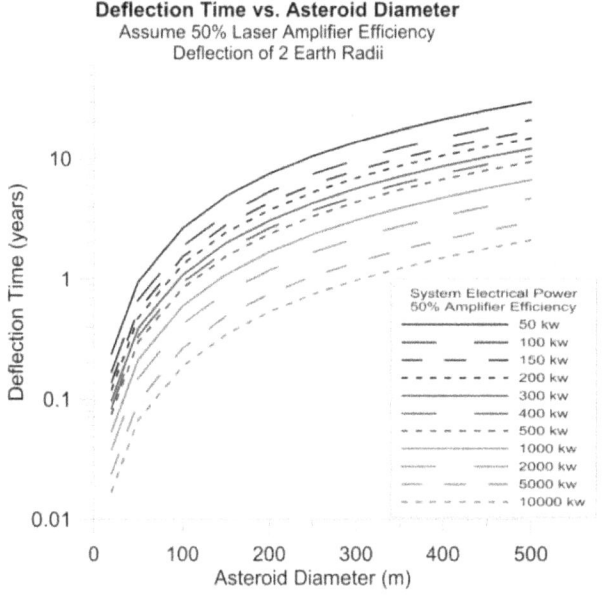

Figure 5.2.3.2 Time to Deflect [Luben, 2015]

Another method that leverages the Apex Anchor for asteroid threat mitigation is focused solar energy. The Apex Anchor provides a platform to extend large concave mirrors to collect solar energy and reflect it towards the offending asteroid [Melosh, 1993] [Nemichnov, 1993] The "mirror" could be a modified solar sail, which the aerospace community has experience in manufacturing and deploying. The authors presented an equation that related the size of the needed mirror to deflect an asteroid of a given diameter within a given timeframe:

$$D(m) = 0.16 \frac{d(km)^{1.5}}{t(yr)},$$

where D is the diameter of the mirror, d is the diameter of the asteroid, and t is the timeframe for deflection. We estimate that an 18-meter diameter mirror could deflect a 500-meter diameter asteroid. A larger mirror can deflect larger asteroids within the same time frame. For example, a 150-meter diameter mirror can deflect a 2-kilometer asteroid in three years.

5.2.4 Space Construction Center

In order for Planet Earth to reach and explore the Universe, there must be a space vehicle large enough for the crew to work as well as sleep. It simply is not feasible to build their large space vehicles on the Earth and they launch from an Earth Space Port. Nor is it feasible for the materials to be launched from individual rockets. The Space Construction Center is designed to be able to build on orbit, such large-scale space vehicles with the materials being delivered via the Tether of the Space Elevator. The Space Construction Center would be located close to the Apex Anchor to allow easy transportation of personnel and materials to/from the Apex Anchor. The Construction Workers would live in the Hotel located at the Apex Anchor and then "commute" to/from the Space Construction Center.

5.2.4.1 Galactic Space Train Construction
As an example, consider constructing the Galactic Space Train. We envision the Galactic Space Train to have a crew complement of about 100 personnel and about 900 colonists. Given this, would have around 500 rooms (For example, kitchens, bedrooms, lounges, and work areas). This would be around 3,710,000 tons (3,364,950 metric tonnes).

5.2.4.2 Delivery Using Rockets
Currently, a single Falcon-6 rocket can essentially inject 18.8 metric tonnes to Space (Mars or Venus) [Falcon Heavy, 2019]. For the sake of this discussion, let's assume each rocket can deliver 18.8 metric tonnes to the Space Construction Center. At 18.8 metric tonnes per rocket, it would take approximately 178,988 launches to deliver that much material to the Apex Anchor. Currently, a single Falcon 9 rocket launch costs approximately $67 million [Urban, 2023]. Given that it would take 178,988 launches, this would cost 12×10^{12} dollars.

Now, let's look at the time to deliver the required 3,365,950 metric-tonnes to the Space Construction Center. The Falcon-6 turnaround time is approximately 30-days [FalconX, 2023]. Therefore, it would take approximately 178,988 months (14,916 years) for the Falcon-6 rockets to deliver the required payload to the Space Construction Center. Naturally, with multiple rockets launches this can be significantly reduced. For example, if four launches per month could be achieved then it would still take 3,729 years to deliver the required materials.

5.2.4.3 Delivery Using the Space Elevator
During the timeframe of 2050, a single Space Elevator Climber can essentially deliver 40.0 metric tonnes to Space and approximately 170,000 metric tonnes per year. For the sake of this discussion, let's assume the Space Elevator System can deliver 170,000 metric tonnes per year to the Space Construction Center [Swan, 2021]. At 170,000 metric tonnes per year, it would take approximately 19.8 years to deliver the materials to the Apex Anchor.

Currently, a single Climber trip costs approximately $220 per Kilogram or $220,000 per metric tonnes. Given that it would take 19.8 years, this would cost 741×10^9 dollars. Naturally, this still seems like a long time, but 19.8 years is far better than 3,729 years by the current Falcon-6 rocket.

Figure 5.3.1: Hospital Ward Aboard the Apex Anchor[12]

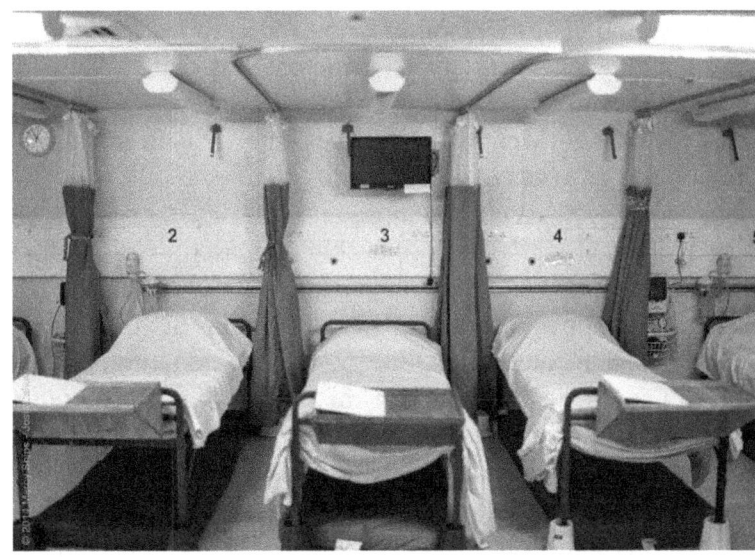

5.3 Space Hospital at the Apex Anchor

Imagine you are working as a scientist performing experiments within a small space station orbiting Earth. For about a week, you have had occasional twinges of pain on the right side of your body, but they quickly pass and you assume they have something to do with the adjustments your body is making while working in microgravity. In the middle of an experiment, that familiar twinge occurs, but this time, it doesn't stop. In fact, it gets worse. Two minutes later, the pain was so bad that you doubled over, calling for one of the other astronauts whose secondary job is Crew Medical Officer (CMO). With the help of voice communications to a flight surgeon on Earth, it is determined that you have acute appendicitis and your appendix is about to rupture. The space station you are in has no medical or surgical facility, just a designated space in a module that serves as a clinic. You need surgery, and you need it within 24 hours to avoid medical complications that are likely to become more severe the longer the surgery is delayed. [Mayo, 2024]

Should an appendix rupture or other medical emergency occur on the International Space Station (ISS) today, NASA and Roscosmos would immediately begin planning a "scoop and run" mission to get the astronaut patient into the Soyuz capsule docked at the ISS for emergency evacuation purposes and returned to Earth for the astronaut patient's emergency surgery. [NASA-ISS, 2024] This may sound like a good, immediate solution, but the reality is that an unplanned return of a Soyuz to Earth takes about 33 hours from notification to touchdown. [Garcia, 2021] Even if the astronaut patient is stabilized prior to transport, the Soyuz is designed for compactness and efficiency, not human comfort, so the ride home will be uncomfortable and the astronaut patient risks further injury due to the nature of the capsule's operations during reentry, landing, and recovery. [Hatfield, 2015]

In a scenario where you are an astronaut on the ISS needing emergency surgery for your appendix that is about to rupture, you are in for an unpleasant experience. Your Crew Medical Officer has had enough medical training to stabilize you and monitor your condition but may not be a medical professional. Receiving guidance from a flight surgeon on Earth, you are administered painkillers and hydration intravenously. [Krishna, 2022]
Several hours later, while the mission control center completes flight planning for your Soyuz capsule's return to Earth, you are helped into a spacesuit, crammed into one of the capsule's upright seats, and the IV is hung near you using zip ties around a conduit, pipe, or other stable

---
[12] This Photo by Unknown Author is licensed under CC BY-SA-NC.

item on the bulkhead. Another IV bag is strapped next to it within your reach to be used later in your flight. Right before the capsule hatch is sealed, the CMO provides you with a crash course in injecting substances into your other IV so that you can self-administer anesthetics after the ones you just got wear off. Then they inject Ketamine into your active IV. As the Soyuz is undocking, you pass out for a while.

You awaken groggily to a noticeable increase in temperature and the capsule is shaking. Reentry into Earth's atmosphere has begun and your appendix painfully reminds you that it is still there and still unhappy. You remember the syringes in your spacesuit's breast pocket and try to decide whether you are alert enough to poke a needle into your second IV's line and squeeze in the relief juice. You can hear a transmission from mission control who, through vital signs monitoring, must have determined that you are awake, but the pain makes it difficult to concentrate and understand what they are saying. You squeeze the transmit button on the armrest and belt out something like, "I'm OK. Going to give myself a shot of morphine." The pain in your side is so bad that you do not hear their response and you feel like you are going to pass out. As the capsule shakes even more, you manage to get the air bubble out of the morphine syringe and inject it into the second IV's line, then open the valve and let the medicine flow into your vein. You leave the Ketamine syringe in your pocket because you decide that you would rather be awake for the journey, especially if it is going to be your last. Several hot, bumpy minutes seem like several hot, bumpy hours and you briefly experience silence in your ears as you pass through the region of reentry where the communications blackout occurs. Gravity is pulling on your body at g's you are not used to, making you feel like you cannot get a full breath due to the elephant sitting on your chest. The morphine you just injected into your IV is no longer working. The pain has returned in full force, but it is too turbulent to inject the Ketamine into your IV line, now. The combination of your body feeling its full weight and then some, and the pain radiating throughout your entire body is too much and you pass out.

Within the next 10 years, several governments and private companies plan to operate up to five new space stations in Earth's orbit. Upgrades to China's Tiangong Space Station are planned with an increase in the number of astronauts aboard [China, 2024], and NASA plans for the continued use of the ISS through 2030 [NASA-FAQ, 2023] [NASA-ISS, 2022]. The resident population of low Earth orbit will increase significantly, which also increases the chances of a medical emergency occurring in space.

A better solution than a long, bumpy return to Earth in a capsule is a much shorter journey to a hospital located at the apex anchor of a space elevator. A hospital at the apex anchor would provide many medical advantages to astronauts working in Earth orbit, lunar orbit, or who are stationed on the Moon, [Donaldson, 2024] and to space tourists who have adapted to microgravity. By diminishing the need to return to Earth for most medical conditions, space workers and tourists would receive more timely medical treatment in a compatible environment from the apex anchor hospital.

Intelligently augmented hospital systems and staff could monitor space workers' health continuously for preventive purposes and alert astronauts and medical care givers to possible adverse medical conditions early.

Surgery in space would be less stressful on a human body adapted to microgravity than transporting the person to Earth, even by space elevator. It takes only a week for the human body to adapt to microgravity, [NASA-ISS-1, 2022] and returning a human to Earth's full gravity for emergency surgery or major medical treatment would add unnecessary stress that could exacerbate a severe medical condition.

While surgery in space is not practiced today, a research and training facility adjacent to the apex anchor hospital would provide a location for the development of new and innovative techniques, procedures, and solutions for treating medical conditions and performing surgeries in space. The benefits of such research would be immediate for space workers and would provide a foundation for long-term medical practice for the future when humans live on the Moon and other planets.

5.3.1 Medical Jobs in Space
The medical industry would see a new discipline arise for medical doctors, surgeons, nurses, researchers, psychiatrists, and facilities maintenance personnel who choose to live and work at the apex anchor hospital. These pioneers would discover best practices, create medical innovations, and establish the protocols and procedures for all future hospitals operating in low and microgravity. They would become the subject matter experts, imparting knowledge, experiences, and wisdom on surgical techniques, environment sterilization, treatment administration, patient transport, convalescence, and a host of associated activities to their Earthly counterparts. Many would educate the next generation of space medical professionals in the forms of hands-on training, simulation, tele-learning, and classroom and laboratory education.

The research facility would provide laboratories to further medical science in space by performing experiments and walking through procedures on simulated subjects prior to performing them on humans. In time, the facility could provide greater laboratory space for the study of long-term space travel on the human body and research ways to counter adverse effects. The benefits of all research at the hospital would be immediate for space workers and would provide a foundation for long-term medical practices for humans working in space for any length of time.

A new dimension of patient care will evolve as a result of humans being treated at an apex anchor hospital in space. Let's return to the scenario where you are an astronaut working aboard a space station who has an appendix that is about to rupture, but this time, there is a hospital at the space elevator's apex anchor. You are doubled over in pain and have called the CMO for help. The CMO contacts the flight surgeon at the apex anchor hospital, and using vital signs monitoring and video telehealth, the flight surgeon diagnoses you with acute appendicitis. It is determined that you need surgery immediately. The CMO provides transitional care, including IV pain killers and hydration, as directed by the flight surgeon. The space station crew works with the mission control center to plan transportation for you in an autonomous ambulance shuttle to the apex anchor hospital. Meanwhile, the flight surgeon at the hospital coordinates with the medical staff to prepare for your arrival.

In about an hour, the CMO escorts you to the autonomous ambulance shuttle. You lie back and are strapped into a padded couch, and the CMO stabilizes you within the vehicle as final flight planning to the apex anchor hospital is completed. Before launch, the CMO provides you with IV morphine and ketamine and closes the shuttle's hatch.

Mission control loads the flight plan into the shuttle's flight system and checks all systems for a "go for launch" indication. You are sedated as the shuttle undocks from the space station and follows its flight plan, speeding up and slowing down to change orbits on its route to the apex anchor hospital. When mission control confirms that the shuttle has successfully docked at the hospital and pressurization is equalized, hospital personnel retrieve you from the shuttle and float you into a chamber for surgical preparation. You are awake enough to answer a few questions and meet the people who will be taking care of you through your emergency surgery. You are in no pain and you feel no stress. Several hours later, you awake in a recovery chamber and are greeted by staff who provide postoperative care for you until you can be transferred to the apex anchor hotel to convalesce.

The plausibility of this scenario depends on the sponsoring government or company's willingness to invest in such a shuttle or spacecraft for emergency medical transportation. Most likely, a multi-purpose vehicle would be used for medical transport between space stations and facilities at the apex anchor. The technology for unpiloted space operations already exists and will be refined further in the coming years, making the technological portion of this scenario plausible. [BAH, 2004]

5.3.2 Rehabilitation Center

The Rehabilitation Center can be used to simulate minimal gravity effects - Can move from center (zero gravity) to outside (max gravity).

Figure 5.3.2.1: Rehabilitation Center Functions

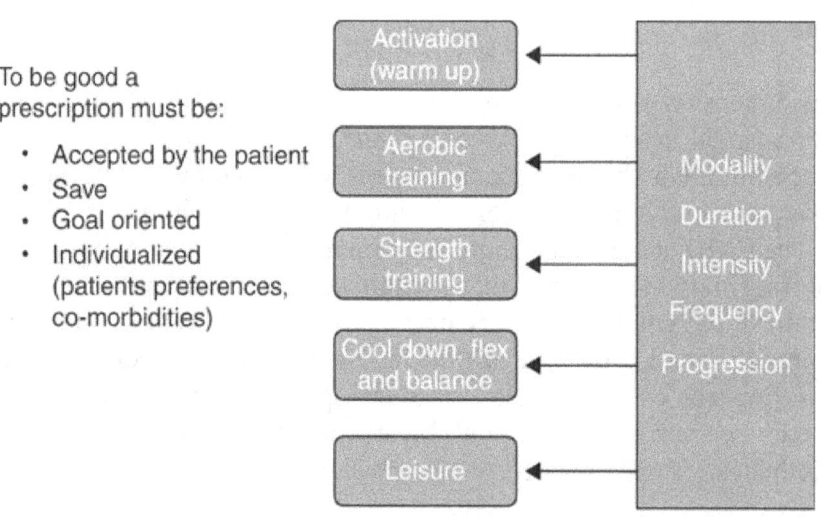

This Photo by Unknown Author is licensed under CC BY-NC-ND

Figure 5.3.3.1: Surgery Center

5.3.3 Surgical Center

"Conducting surgery in space is not something we've had to contend with yet, but the further we travel from Earth, the more likely it is that we will need expanded medical capabilities," said George Pantalos, Ph.D., principal investigator for the SFMS at University of Louisville. "And because of the microgravity environment, surgery and wound care in space will be very challenging."

In the absence of gravity, blood droplets and other fluids from wounds or surgical sites can float into the cabin of a spacecraft, contaminating equipment and potentially introducing disease. The SFMS includes a clear dome that creates a tight seal to a patient's skin and provides insertion points for surgical instruments without letting fluids escape. A multi-function surgical device (MFSD) performs several tasks in one wand-like instrument: suction, irrigation, illumination, vision, and cautery.

"For long-duration human spaceflight missions to the Moon, Mars, and other destinations, there is a need to monitor the state of astronaut health and, when necessary, make the appropriate interventions in response to health changes or the onset of disease," said Richard Mathies, Ph.D., principal investigator for the Lab-on-a-Chip.

Large labs and equipment required for conducting clinical analyses here on Earth are not practical in a spaceflight setting. Instead, astronauts rely on smaller systems like WetLab-2 (currently in operation on the International Space Station) to monitor signals of declining health. As small lab systems become even more miniaturized and fully autonomous, they may be applicable to even more space-based uses, such as sample analysis conducted as part of investigations for signs of life on other planets and moons.

"These suborbital parabolic flights are critical to demonstrating the Lab-on-a-Chip capabilities in a challenging zero gravity environment where fluids behave very differently," said Anna Butterworth, Ph.D., at the University of California – Berkeley and Jungkyu Kim, Ph.D., at the University of Utah, co-investigators for the project. "This is a key step to successfully developing dedicated microfluidic systems for clinical health monitoring in space."

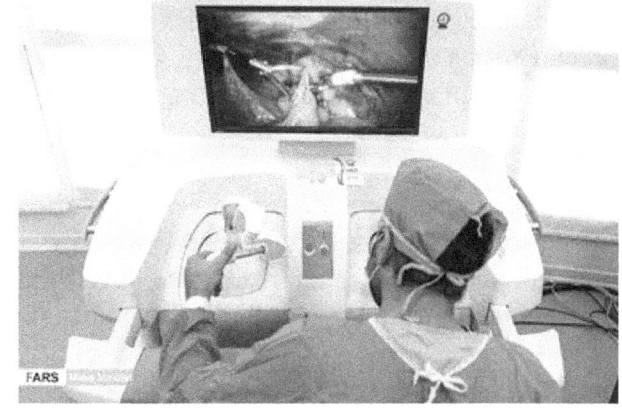

Figure 5.3.3.2: Remote Surgery

Using current technology, surgeries can be conducted remotely as shown in Figure 4.5.3.1.

Figure 5.4.1: Hotel Lobby at Apex Anchor[13]

5.4 Space Hotel at the Apex Anchor

Living and working in space changes a person's physiology [Faini, 2021]. After only a few months, the human body adapts to microgravity, muscles atrophy even with exercise as the resistance that keeps them robust does not exist in space. Therefore, traveling to Earth after an extended work assignment in space can be very hard on even a healthy human body. Astronauts returning to Earth after six months on the International Space Station typically take two to three days to readjust to full gravity, and eyesight adjustments and vertigo can take longer to remedy themselves. And they're in really good shape.

Imagine that you are an astronaut scientist working aboard the ISS on a year-long assignment. A few months ago, a flawed machine put your work behind schedule and you have been working extended hours with no days off for several weeks to get the experiment back on track, and as a result, you are feeling lethargic and unhappy. Burnout has set in. What if you could take an autonomous shuttle from the ISS to a hotel at the apex anchor of the space elevator to get in some recreation and recharge? This would be a less expensive and more timely option than sending you to Earth where it would take you a few days to physically recover from the trip before your recreation can begin, and then having to wait until the next rocket mission to the ISS to return you to your valuable work.

A space hotel at the apex anchor would provide large spaces for socializing, sports, music, dining, entertainment, classes, conferences, events, weddings, celestial and Earth observation, and supervised extravehicular activities (EVA). Think of how hilarious a game of dodgeball would be in microgravity, using little blips of compressed air to propel yourself around a room. Better not to use too much or you risk running out and becoming stranded and an easy floating target. Classes in painting, sculpture, astronomy, cooking, cocktail making, and other interesting activities would be offered by visiting artists and experts for a challenging, different take on hobbies.

A variety of unique entertainment would be offered to hotel guests. How different would a string quartet sound? How much greater and spectacular would an acrobatic troupe perform in microgravity? What would dancing be like? What would sex be like? All of these activities could be explored in depth in spaces built for human movement and activity, with padding and hook-

---

[13] This Photo by Unknown Author is licensed under CC BY-SA.

and-loop fasteners on the walls and your suit to prevent harm and help arrest vigorous momentum. As an astronaut escaping burnout, just a few days of frivolity in a location to which you are already acclimatized could be the perfect remedy. The turnaround time would be shorter; there would be no need to wait for adjustment to full gravity or for the next expensive rocket headed to the ISS to return to your work. You could be absent a few days instead of a few months with the same benefit.

The hotel at the apex anchor could provide more activities and benefits to the human race than entertainment. The opportunities for education, environmentalism, and artistic exploration are astronomical (pun intended). A field trip of a few days with primary school aged children would be the educational experience of a lifetime. Staying at the hotel would mean experiencing microgravity at flexible ages, giving some children with physical disabilities experiences with movement they could never have on Earth. Imagine a room full of hammocks secured to the walls by industrial hook-and-loop fasteners with snoozing children in them like caterpillars waiting to emerge. The children would be worn out from the day's activities of bouncing off similarly padded walls, engaging in physical games, inventing new and more efficient ways to move and do things in microgravity, performing educational activities, and looking out of observatory windows and expanding their minds into the vastness of space.

Looking out of those same windows, people of all ages and education levels could explore outside and within, seeing the green, blue, and brown Earth below, which often results in the Overview Effect (2), where great appreciation is gained for the uniqueness and fragility of our home planet. Opportunities for professional and amateur astronomy abound through those same windows and with an observatory installed in one segment of the hotel. Findings and discoveries could be documented and elaborated upon in secluded rooms set aside for quiet research, writing, and the creation of art.

Beyond elementary education, a space hotel with conference facilities would be an excellent venue for college-level classes in science, philosophy, astronomy, medicine, and innumerable other topics, with accompanying labs in specialized hotel spaces. Conference rooms could be provided to allow attendees from many industries to experience first-hand the benefits and challenges of living and operating in space, thereby making informed decisions for government agencies and corporations.

The hotel could also be a place of convalescence for space workers recovering from major surgeries or illness. As discussed earlier in this document, the likelihood of major surgeries or treatments for space workers will increase as the number of people working in space increases. It will not always be practical or healthy for such patients to return to Earth immediately after treatment. A space hospital will have limited beds for recovery after a major procedure and sometimes people need to rest and recuperate for longer periods of time and receive physical therapy. The hotel at the apex anchor would offer comfortable rooms for convalescence with nursing staff and at least one physician and physical therapist already on board who could provide necessary care. After discharge from the hospital, the patient could be transported by floating them down a corridor connecting the hospital to the hotel, through a pressurization chamber, and into the hotel receiving area where the patient would be greeted by a nurse and hotel staff for check-in. The patient would then be floated to their convalescence room, monitored, and cared for until ready to return to their place of work. The social and entertainment aspects of the hotel environment would be beneficial for patient morale, as would the ability of family members to take a trip up the space elevator to visit their loved one at the apex anchor hotel. Upon notification that a loved one is severely injured in space where recovery is unlikely, loved ones and perhaps the patient's religious leader or minister could make the trip up the space elevator, stay at the hotel, and be at the hospital by their loved one's side during the patient's transition from life, instead of being impossibly far away during such an unfortunate family event.

For those needing more help adapting to microgravity or who wish to reacclimate to the standard gravity of Earth before returning, a spinning, spoked ring emanating from a hub on the hotel would allow people and animals to experience and adjust to varying degrees of gravity. Secluded rooms for people wishing to avoid sleeping in microgravity could be built at the farthest ends of the ring's spokes, providing simulated gravitational force. Similarly, rooms near the center of spokes could provide comfortable accommodations to people with physical challenges, allowing them more freedom of movement than 1 g but with enough gravity to move about without free-floating. These rooms could also provide spaces for physical therapy, providing less stress on the body.

Research areas along the spokes and inside the ring would be available for rent or lease to scientists, educators, artists, and students to study and research the effects of various degrees of gravity on plants, animals, other living things, viscosity, and materials. The spinning ring can also be a source of agricultural studies for long duration space missions and also provide fresh food for the hotel and hospital kitchens. Waste generated by the hospital and hotel can be sterilized and composted within the ring structure to provide soil and nutrients for plants and animals living and growing in the hotel ring.

Keeping such a facility operating smoothly would require a wide range of skill sets and personnel, resulting in jobs unique to space facility upkeep and hospitality. The ability to clean and sanitize spaces, materials, and textiles will require a staff of environmental and facilities specialists, many of whom will be developing methods that will be used on long space journeys. Keeping environmental equipment and machines functioning, the ring operating and maintained, and facility repairs will require engineers and technicians who are able to function well in microgravity and are trained and experienced with EVAs for external facility repairs. Agriculture specialists and horticulturalists will be necessary to maintain plants and crops in the ring and would provide valuable data to scientists regarding plant growth, sunlight needed, water use, and other parameters. Hospitality personnel will need to be versed in more than visitor care and needs; they will need to be proficient in safety procedures, security, egress, and emergency procedures, similar to those skills needed by such staff aboard a cruise ship. Chefs will be challenged to create dishes that have enough flavor to satisfy guests looking for unique experiences, will need to find new methods to heat and cook food, and will need to consider the changes in olfactory and taste senses that occur in humans in space. Event planners and administrative personnel will have their hands full with providing unique and varied activities for hotel guests. Experienced, retired astronomers and astronauts will be on staff to provide guided observations of the universe and be able to supervise tethered EVA excursions out of the hotel and into space. Visiting artists and performers will require the setting up of specialized equipment and area accommodation for performances. A medical staff will need to be in place to help guests with vertigo, adjusting to microgravity, bumps and scrapes from activities, and checking in on convalescing patients staying at the hotel.

Several companies have already considered building hotels in low Earth orbit [Mafi, 2023], and have conducted research into creating such facilities in the near future. Leveraging their discoveries, it would be possible to create a stationery hotel in a matter of a few years after apex anchor construction, with the spinning spoked ring coming up to speed a few years after that. The advent of inflatable space habitats to supplement permanent structures during construction makes a hotel at the apex anchor a quick reality [Matthewson, 2021].

Support to tourism (and humans at the Apex Anchor) would require a maturation of capabilities and will be planned for the fully operational capability estimated to be ten years after initial operations. This would include missions to the Moon and Mars as well as tourism to anywhere in the area. The concept is that a large human rated space system would be assembled and fueled at the Apex Anchor so that humans can raise on tether climbers and then release towards destinations such as around the Moon, landing on the Moon and then on to Mars.

5.4.1 Hotel Jobs in Space
The hotel industry would see a new area of growth when dealing with a hotel at the Apex Anchor. Consider all the disciplines required to run a hotel – maids, cooks, maintenance, laundry, just to name a few. Consider those who choose to live and work at the apex anchor hotel. Many would educate the next generation of space hotel workers on how to live and work in space.

Chapter 6: Study Summary, Conclusions and Recommendations

6.1 Summary

The obvious emphasis from the above discussion is that the descriptive word is transformational. The implementation of Space Elevator Transportation Systems around the world will enable remarkable capabilities to move cargo massively and routinely. This will transform the future of space activities. Space elevator leadership must point out and emphasize to both the general public and project managers of visionary space activities that the modern-day space elevator will be:

- Transformational
- A partner with rockets in a Dual Space Access Architecture
- Green: lifts payloads with electricity and enables Space Solar Power satellites to GEO with a timely schedule
- Closer than they think as SETS has entered engineering development
- A program for timely delivery of massive loads for development
- An economic boom as regions opens up for commerce because of their routine, daily, and massive lift capabilities.

As discussed in Chapter's four and five, it is envisioned that the Apex Anchor will be able to host a variety of space missions, such as:

Apex Anchor Operations Center
Mission: Support all customers in day-to-day operations

Planetary Defense
Mission: Protect the planet

Space Transportation Port (A.K.S. "Truck Stop")
Mission: Provide facilities to accept space systems entering the Apex Anchor [from orbits or from the tether], service the vehicles, and release back to orbit or along the tether

Space Construction Platform
Mission: Support and enable construction of facilities and space systems with ability to build, repair, improve upon the Apex Anchor

Space Hospital and Rehabilitation Center
Mission: Attend to the medical needs of the residents and the transient personnel upon need.

Space Logistics, Storage and Distributing Center
Mission: Enable the operations of multiple missions upon the Apex Anchor with storage and distribution of logistics.

Space Solar Power Distribution Center
Mission: Enable the collection of solar energy and then the ability to distribute around the Apex Anchor and to regional space systems in need of electrical energy.

Space Astronomical Observatory
Mission: Enable quiescent location for observation of the arena around the Apex Anchor and the solar system and universe as needed.

Nuclear Waste Disposal System
Mission: Allow nuclear waste from the Earth to transit the Apex Anchor on its way from the tether to a destination (orbit) far away from the Earth for terminal storage of high-level nuclear waste

Space Hotel
Mission: Enable transients and residents to have living quarters within the Apex Anchor that are safe and convenient for the work environment.

Next Generation International Space Station
Mission: Empower multiple countries to participate in a space station at the Apex Anchor that would enable a stable environment for government & commercial operations and existence.

6.2 Conclusions

The space elevator offers much more than rockets; it is not just a way to travel into space. Transportation infrastructure is a better description; like the tracks for trains, it will open the way to an unexplored space for transporting people and also heavy loads, but in the case of space elevators, that space is far above the Earth rather than into America's "Wild West." It will allow settlements to be developed at its apex and to the moon and other planets by transporting materials too heavy for rockets. Payloads will include equipment to be used for missions such as the creation of planetary defense systems, nuclear waste disposal systems, astronomical observatories, even hotels and a hospital for travelers and workers to get care equivalent to that which is found on Earth.

Space Elevators have many strengths; but the most remarkable ones relate to their permanent space transportation infrastructures. They will move massive cargo daily, routinely, safely, and environmentally friendly. The potential missions aboard the Apex Anchor, could be viewed as the potential to be a "waystation" for settlements to the cosmos.

To achieve maximum benefit from the high speed of the apex anchor around the earth, it is best to introduce motion at right angles so that all directions in the celestial sphere can be reached. This has the added benefit that velocities in some directions can be as high as 17.7km/s. Table 6.2.1 provides some examples of velocities vs. distance from the Earth. Lengthening the tether allows even higher velocities with Spacecraft released.

Table 6.2.1: Velocities vs. Distance from the Earth[14]

Distance from the Earth (Km)	Relative Velocity (Km/s)
100,000	7.76
130,000	9.94
150,000	11.4
163,000	12.35

The unique and powerful concept of basing a Planetary Defense System at Apex Anchors is simple and achievable. The Concept is:
- Identify near term threatening asteroids (i.e., coming out of the Sun) with stereoscopic observations, and
- Rapid assembly [segments stored at the Apex Anchor] and release of Planetary Defense "asteroid busters" space systems within 24 hours towards rapidly approaching threats.

Figure 6.2.1: Apex Anchor "Garage"[15]

As mentioned earlier, the Apex Anchor Planetary Defense system could be "housed" within a "Garage" like structure within the Apex Anchor. Figure 6.2.1 is a conceptual picture of this "garage" for the Planetary Defense system aboard the Apex Anchor. This approach enables reaching any destination in the solar system at high speed with fuel only needed for course correction and deceleration.

Currently, a single Falcon-6 rocket can essentially inject 18.8 metric tonnes to Space (Mars or Venus). If there were three launches per day for 365 days this would equate to 20,586 metric tonnes. [Falcon Heavy 2019]. However, the Space Elevator at IOC can deliver 30,000 metric tonnes and at FOC can deliver 170,000 metric tonnes per year.

Doors to many jobs for adventurous people will be open to fulfill these roles in space. In the *Space Elevator Near Term Vision* section readers were asked to imagine being on a trip into space. Now we ask you to go beyond that image and think of what part you could play after the Space Train has brought thousands of travelers to the Apex Anchor, some to stay. Imagine how you could share your expertise and earn a good living either in commercial jobs or in some official capacity, maybe while living in a Martian colony. There will surely be a place for engineers, hospitality and health care workers, construction personnel, even astronomers, climate entrepreneurs and business and IT specialists. And of course, scientists, researchers' teachers

---

[14] Method of Computation: Distance from Earth Center (Km) x 7.2921E-5/s.
[15] This Photo by Unknown Author is licensed under CC BY-SA

and writers (to publicize ways to join the growing numbers who want to make a difference and believe they can do that most effectively in space, either remotely or physically).
If you have caught the vision of the benefits of space elevators for our future, please go to ISEC.org and study the reports or offer recommendations for future missions.

Final Conclusion: This ISEC research report developed the concept of assembly, refuel, repair or full construction of large space systems above the gravity well. This concept is unique as each of the segments arrives at the Apex Anchor with tremendous energy gained by rising from the ocean surface. In essence, each segment is similar to a fully fueled segment of a truck convoy resulting in a Full-Service Transportation Node. The concept, illustrates this simple idea – let's build huge space systems that already have monumental energy stored by being out of the gravity well and having huge velocities upon release. The study continued to develop the concept and came up with a remarkable number of missions that would be revolutionary in concept, but evolutionary in movement off-planet.

6.3 Recommendations

Initiate research into the future missions of importance for humanity's dreams; then discuss methods for Apex Anchors to contribute towards those missions.

References

- [Alberts, 2001] Albers, David S, Garstka, John J., Hayes Richard E., Signori, David, "Understanding Information Age Warfare", CCRP, 2001, p. 3.

- [Alberts, 2001] Albers, David S, Garstka, John J., Hayes Richard E., Signori, David, "Understanding Information Age Warfare", CCRP, 2001, ISBN 1-893723-04-6

- [Alberts, 2001] Albers, David S, Garstka, John J., Hayes Richard E., Signori, David, "Understanding Information Age Warfare", CCRP, 2001, p. 57-58.

- [Alberts, 2001] Albers, David S, Garstka, John J., Hayes Richard E., Signori, David, "Understanding Information Age Warfare", CCRP, 2001, p. 146.

- [Alberts, 2003] Alberts, David S, Hayes, Richard E., "Power to the Edge: Command…Control… in the Information Age,", Information Age Transformation Series, CCRP, 2003, p. 100.

- [Alberts, 2011] Alberts, David S., "The Agility Advantage: A Survival Guide for Complex Enterprises and Endeavors," CCRP, ISBN 978-1-893723-23-8, 2011, p. 3.

- [Barry, 2021] Barry, K, Alfaro, E.P., "Changing the Economic Paradigm for Building a Space Elevator," 72nd International Astronautical Congress, 2021

- [BAH, 2004] Booz Allen Hamilton, Future Interagency Range and Spaceport Technologies (FIRST). Space Vehicle Operators Concept of Operations. October 2004. <https://www.faa.gov/about/office_org/headquarters_offices/ast/media/Space_Vehicle_Operators_CONOPS_v18.pdf>

- [Bozzo, 2024] Bozzo, E., "Future Prospects for Gamma-ray Burst Detection from Space," 2024, Universe, 4, 187

- [Chandra X-ray Telescope, 2024] O'Callaghan, J., "Astronomers Fight to Save X-ray Telescope as NASA Dishes Out Budget Cuts," 2024, https://www.scientificamerican.com/article/nasa-slashes-budget-for-chandra-its-greatest-x-ray-observatory/, accessed on June 18, 2024

- [Chapman, 2001] Chapman, Durda, and Gold (2001: Wayback Machine (archive.org))

- [Cheng, 2023] Cheng et al. 2023; https://www.nature.com/articles/s41586-023-05878-z

- [China, 2024] China Manned Space. Accessed 2024-05-18. <http://en.cmse.gov.cn/aboutcms/>

- [Combes, 2023] Combes, F., "Galactic Bulge-Black hole Co-evolution, feeding, and outflows in AGN," 2023, https://arxiv.org/pdf/2302.12917, accessed on June 18, 2024

- [Donaldson, 2024] Donaldson, Abbey A. NASA Shares Progress Toward Early Artemis Moon Missions with Crew. 2024-01-09. <https://www.nasa.gov/news-release/nasa-shares-progress-toward-early-artemis-moon-missions-with-crew/#:~:text=NASA%20will%20now%20target%20September,remains%20on%20track%20for%202028.>

- [Eddy, 2021] Eddy. J. etc., "Space Elevators: The Green Road to Space," ISEC Study Report, www.isec.org, 2021.

- [Eddy, 2023] Eddy, J. etc. "Leverage Dual Space Access Architecture Advanced Rockets and Space Elevators, ISEC Study Report, www.isec.org, 2023.

- [Faini, 2021] Pantalone D, Faini GS, Cialdai F, Sereni E, Bacci S, Bani D, Bernini M, Pratesi C, Stefàno P, Orzalesi L, Balsamo M, Zolesi V, Monici M. Robot-assisted surgery in space: pros and cons. A review from the surgeon's point of view. NPJ Microgravity. 2021 Dec 21;7(1):56. doi: 10.1038/s41526-021-00183-3. PMID: 34934056; PMCID: PMC8692617. <https://www.ncbi.nlm.nih.gov/pmc/articles/PMC8692617/>

- [Falcon Heavy, 2019] Technical data sheet, Falcon Heavy, No. 51, June 2019, pp. 62-63.

- [FalconX, 2023] Falcon 6 launched 23 starlink broadband satellites in 2023. Therefore, assume one per month.

- [Fitzgerald, 2017] Fitzgerald, M etc., "Design Considerations for the Space Elevator Apex Anchor and GEO Node," ISEC Study Report, www.isec.org, 2017

- [Fitzgerald, 2020] Fitzgerald, M etc., "Space Elevators are the Transportation Story of the 21st Century," ISEC Study Report, www.isec.org, 2020.

- [Garcia, 2021] Garcia, Mark A. NASA's SpaceX Crew Rescue and Recovery. 2021 April 19. Retrieved from the following on 2024-05-11 <https://www.nasa.gov/humans-in-space/nasas-spacex-crew-rescue-and-recovery/>

- [Gauri, 2024] Gauri, S., et al. "Tully-Fisher Relation of late-type Galaxies at $0.6<z<2.5$," 2024, https://arxiv.org/pdf/2406.08934, accessed on June 18, 2024

- [Hatfield, 2015] Hadfield, Chris. An Astronaut's Guide to Life on Earth: What Going to Space Taught Me About Ingenuity, Determination, and Being Prepared for Anything. 2015 April 14. ISBN-13 978-0316253031.

- [ISEC Website, 2024] This section is from What is a Space Elevator on ISEC website www.isec.org 10 Feb 2024.

- [ISEC Webpage, 2024] Lead into International Space Elevator Consortium home page at www.isec.org. (10 Feb 2024)

- [Johnson, 2016] Johnson, Lindley, Planetary Defense Coordination Office Update, 2016

- [Kempton, 2024] Kempton, E. M. -R., & Knuston, H. A., "Transit Exoplanet Atmospheres in the JWST Era," 2024, https://arxiv.org/pdf/2404.15430, accessed on June 18, 2024

- [Klessen, 2023] Klessen, R. S., & Glover, S. C. O., "The First Stars: Formation, Properties, and Impact," 2023, Annual Reviews of Astronomy & Astrophysics, 61, 56

- [Knapman, 2022] J.M. Knapman, P.A. Swan, Secondary tethers, **Acta Astronautica**, 195 (2022) 561-566

- [Knapman, 2023], Innovation and research for space elevators, **Journal of the British Interplanetary Society**, 76 (2023), 190-196

- [Krishna, 2022] Krishna, Swampna. How do Astronauts Handle Medical Emergencies in Space? 2022-11-11. <https://now.northropgrumman.com/how-do-astronauts-handle-medical-emergencies-in-space>

- [Lubin, 2015] Lubin et al, ; https://www.deepspace.ucsb.edu/wp-content/uploads/2013/09/PDC-2015-Lubin-e.pdf

- [Mafi, 2023] Nick Mafi, Katherine McLaughlin. Architectural Digest, A Hotel in Space Could Be Operational in Just Five Years. <https://www.architecturaldigest.com/story/worlds-first-hotel-in-space>

- [Matthewson, 2021] Matthewson, Samantha. Space.com, BEAM Inflatable Space Habitat Has Successful 1st Year in Orbit. <https://www.space.com/37068-beam-inflatable-habitat-first-year-space.html>

- [Mayo-2024] Mayo Clinic, mayoclinic.org. Accessed 2024-05-18. <https://www.mayoclinic.org/diseases-conditions/appendicitis/diagnosis-treatment/drc-20369549>

- [Melosh, 1993] Melosh, Solar asteroid diversion | Nature)

- [Musk 2017] Musk, Elon, "Making Humans a Multi-Planetary Species", New Space, Vol 2, No. 2, 2017.

- [NASA, 2019] NASA. Houston, we have a Podcast. Season 1, Episode 7, 30 August 2019. <https://www.nasa.gov/podcasts/houston-we-have-a-podcast/the-overview-effect/>

- [NASA, 2020] https://'www.nasa.gov/Planetary Defense Coordination Office | NASA

- [NASA-ISS, 2022] NASA Provides Updated International Space Station Transition Plan. NASA. 2022-01-31. Accessed on 2024-05-18 <https://www.nasa.gov/humans-in-space/nasa-provides-updated-international-space-station-transition-plan/>
- [NASA-FAQ, 2023] Frequently Asked Questions About the International Space Station Transition Plan. NASA. 2023-12-13. <https://www.nasa.gov/faqs-the-international-space-station-transition-plan/>
- [NASA-ISS 2024] NASA. Station Overview. Retrieved from the following on 2024-05-11 <https://www.nasa.gov/international-space-station/space-station-overview/>
- [NASA-ISS, 2024] NASA Provides Updated International Space Station Transition Plan. NASA. 2022-01-31. Accessed on 2024-05-18 <https://www.nasa.gov/humans-in-space/nasa-provides-updated-international-space-station-transition-plan/>
- [Naumann, 2023] Naumann, R., Answer to question [What would a space elevator look like from the ground?]. Quora, 27 Apr 2023, . https://qr.ae/psB4yQ
- [Nemichnov, 1993]; Nemichnov, Solar asteroid diversion | Nature)
- [Peet, 2021] M. Peet, The orbital mechanics of space elevator launch systems, **Acta Astronautica**, 179 (2021) 153-171
- [Phister, 2003] Phister, P.W., Jr., Swan, P. A, "Briefing: An Emerging Theory of Warfare: Network Centric Warfare," 2003.
- [Phister, 2021] Phister, P.W., Jr., Swan, P. A., "An Interplanetary Transportation system," Room Space Journal of Asgardia, Issue 4(30) 2021.
- [Pope John Paul II] information at: https://www.goodreads.com/author/quotes/6473881.Pope_John_Paul_II
- [Swan, 2013] P.A. Swan, D.I. Raitt, C.W. Swan, R.E. Penny, J.M. Knapman, "Space Elevators: An Assessment of the Technological Feasibility and the Way Forward", International Academy of Astronautics, 2013
- [Swan 2020] Swan, Peter, "Chapter 1 – Introduction – Opening Interplanetary Missions," Mar20
- [Swan, 2020b] Swan, P. etc. "Today's Space Elevator," ISEC Study Report, www.isec.org, 2020.
- [Swan 2021] "Appropriate Space Access Architecture for Mars: Leveraging Rockets and Space Elevators, 2020.

- [Swan, 2021] Swan "Dual Space Access Strategy Minimizes the Rocket Equation," Space Renaissance International 3rd World Congress 2021 – Congress Theses, Final Resolution and Papers. Pg 254-255.
- [Swan, 2022] Swan, Peter, Lecture at the ASU Osher Lifelong Learning Institute, fall of 2022 entitled "Leave the Rockets – Take a Space Elevator!"

- [Swan, 2023] Swan, PA., Swan, C.W. *Journal of The British Interplanetary Society* • V 76. (2023) Pgs. 247-251, DOI https://doi.org/10.59332/jbis-076-07-0247

- [Urban 2023] "How Much Does It Cost to Launch a Rocket?", Exclusives, Insights, 12Oct23.

- [Woods, 2023] Woods, A., Vogler, A., Collins, C., The Lunar space elevator: A key technology for realizing the greater earth lunar power station, *Journal of the British Interplanetary Society. V. 76. (2023) Pgs. 253-260, https://doi.org/10.59332/jbis-076-07-0252*

- [Wright, 2023] Wright, D. "Building the Space Elevator Tether", Special Issue: Future Directions for Space Elevators", Journal of the British Interplanetary Society Volume 76 No.7 July 2023.

Appendix A: Glossary

Term	Definition
Apex Anchor	Apex Anchor is a Full-Service Transportation node at the top of a Space Elevator.
AAPDS	Apex Anchor Planetary Defense System
AAAO	Apex Anchor Astronomical Observatory
A.U.	Astronomical Units
CISLunar	Indicates something is located between the Earth and the moon, or within the moon's orbit. CISLunar is the region beyond Earth's geosynchronous orbit (GEO) that is influenced by the gravitational forces of both the Earth and the Moon.
DART	NASA's program tilted "Double Asteroid Redirection Test" launched on 24Nov21 and successfully impacted its asteroid target on the evening of 26Sep24.
Earth Port	The Earth surface transportation node anchoring the Space Elevator and providing the transfer capabilities from Earth Port along the tether.
Full-Service Node	A location able to service space systems such as: refuel; repair, assemble, construct, and store with high-speed release towards mission destinations, or capture after return.
Ecliptic Plane	The ecliptic is the plane of Earth's orbit around the Sun. It extends beyond that to include the seven other planets — and, because it's imaginary, actually beyond that into infinity.
FOC	Final Operational Capacity, where the Space Elevator will be capable of delivering 170,000 tonnes per year.
Galactic Harbour	The combination of the Space Elevator Transportation System and the Space Elevator Enterprise System. The Galactic Harbour will be the volume encompassing the Earth Port while stretching up in a cylindrical shape to include two Space Elevator tethers outwards beyond the Apex Anchor.
Gravity Well	The word 'well' is used metaphorically *to account for the pull of gravity that a large body in space exerts.* The larger the body (the more mass) the more of a gravity well it has. For example, Mars is smaller and less massive than Earth, so Mars' gravity well is shallower than Earth's gravity well; the Moon is even less massive than Mars, so the Moon's gravity well is much shallower than either Earth's or Mars' gravity wells
GEO Region	The tether transitions through this region where operational satellites could operate next to the tether [GEO is geosynchronous Earth orbit]
Green Road to Space	Space Elevators are Big Green Machines. They are the Green Road to Space because they enable humanity's most important missions, such as Space Solar Power to improve the Earth's environment with electricity from space. They move massive tonnage to GEO and beyond safely, routinely, inexpensively, and daily while providing a zero-carbon footprint during operations. Their payloads are raised to destinations with electricity (vs. rocket fuel burning in the atmosphere) and they leave no residual components in orbit during operations. In addition, Space Elevators enable Earth friendly missions that cannot reasonably be accomplished with rockets: electricity from space eliminating hundreds of coals burning plants, Earth sunshade to reduce solar energy impact, and construction of massive structures at GEO, Moon and Mars.
LDSAA	Leverage Dual Space Access Architecture

IOC	Initial Operational Capacity, where the Space Elevator will be capable of delivering 30,000 tonnes per year.
MFSD	Multi-Functional Surgical Device – one wand-like instrument that performs several tasks including suction, irrigation, illumination, vision and cautery.
PDCO	Planetary Defense Coordination Office
SETS	Space Elevator Transportation System
Space Elevator	The Modern-Day Space Elevator is a permanent space transportation node reaching from the equator to approximately 100,000 km altitude [Earth based]
TFA	Time-for-Arrival
Tether	The space elevator tether reaches from the ocean surface [or land based] to the Apex Anchor. The material and characteristics are to be designed.
Tether Climber	The movable infrastructure traveling along the tether

Appendix B: International Space Elevator Consortium

Who We Are

The International Space Elevator Consortium (ISEC www.isec.org) is composed of individuals and organizations from around the world who share a vision of humanity in space.

Our Vision

Space Elevators are the Green Road to Space while they enable humanity's most important missions by moving massive tonnage to GEO and beyond. This is accomplished safely, routinely, inexpensively, daily, and they are environmentally neutral.

Strategic Approach: Dual Space Access Architecture

Rockets to open up the Moon and Mars with Space Elevators to supply and grow the colonies. In addition, Space Elevators will enable Green Missions such as Space Solar Power and L-1 Sunshade. This compatible and complementary approach with future rockets is not competitive while leveraging the strengths of both inside a Dual Space Access Architecture.

Our Mission

The ISEC promotes the development, construction and operation of a space elevator infrastructure as a revolutionary and efficient way to make space for all humanity.

What We Do

- Provide technical leadership promoting development, construction, and operation of space elevator infrastructures.
- Become the "go to" organization for all things space elevator.
- Energize and stimulate the public and the space community to support a space elevator for low-cost access to space.
- Stimulate science, technology, engineering, and mathematics (STEM) educational activities while supporting educational gatherings, meetings, workshops, classes, and other similar events to carry out this mission.

A Brief History of ISEC

The idea for an organization like ISEC had been discussed for years, but it wasn't until the Space Elevator Conference in Redmond, Washington, in July of 2008, that things became serious. Interest and enthusiasm for a space elevator had reached an all-time peak and, with Space Elevator conferences upcoming in both Europe and Japan, it was felt that this was the time to formalize an international organization. An initial set of directors and officers were elected and they immediately began the difficult task of unifying the disparate efforts of space elevator supporters worldwide. ISEC's first Strategic Plan was adopted in January of 2010 and it is now the driving force behind ISEC's efforts. This Strategic Plan calls for adopting a yearly theme to focus ISEC activities. Because of our common goals and hopes for the future of mankind off--planet, ISEC became an Affiliate of the National Space Society in August of 2013. In addition, ISEC works closely with the Japanese Space Elevator Association.

Our Approach

ISEC's activities are pushing the concept of space elevators forward. These cross all disciplines and encourage people from around the world to participate. The following activities are being accomplished in parallel:

- Yearly conference – International space elevator conferences were initiated by Dr. Brad Edwards in the Seattle area in 2002. Follow-on conferences were in Santa Fe (2003), Washington DC (2004), Albuquerque (2005/6 –smaller sessions), and Seattle (2008 to the present). Each of these conferences had multiple discussions across the whole arena of space elevators with remarkable concepts and presentations.
- Yearlong technical studies – ISEC sponsors research into a focused topic each year to ensure progress in a discipline within the space elevator project. The first such study was conducted in 2010 to evaluate the threat of space debris. The products from these studies are reports that are published to document progress in the development of space elevators. They can be downloaded at www.isec.org.
- International Cooperation – ISEC supports many activities around the globe to ensure that space elevators keep progressing towards a developmental program. International activities include coordinating with the two other major societies focusing on space elevators: the Japanese Space Elevator Association and EuroSpaceward. In addition, ISEC supports symposia and presentations at the International Academy of Astronautics and the International Astronautical Federation Congress each year.
- Publications – ISEC publishes a monthly e-Newsletter, its yearly study reports and an annual technical journal [CLIMB] to help spread information about space elevators. In addition, there is a magazine filled with space elevator literature called Via Ad Astra.
- Reference material – ISEC is building a Space Elevator Library, including a reference database of Space Elevator related papers and publications. (see section before this on references)
- Outreach – People need to be made aware of the idea of a space elevator. Our outreach activity is responsible for providing the blueprint to reach societal, governmental, educational, and media institutions and expose them to the benefits of space elevators. ISEC members are readily available to speak at conferences and other public events in support of the space elevator. In addition to our monthly e-Newsletter, we are also on Facebook, Linked In, and Twitter.
- Legal – The space elevator is going to break new legal ground. Existing space treaties may need to be amended. New treaties may be needed. International cooperation must be sought. Insurability will be a requirement. Legal activities encompass the legal environment of a space elevator - international maritime, air, and space law. Also, there will be interest within intellectual property, liability, and commerce law. Starting to work on the legal foundation well in advance will result in a more rational product.
- History Committee – ISEC supports a small group of volunteers to document the history of space elevators. The committee's purpose is to provide insight into the progress being achieved currently and over the last century.
- Research Committee – ISEC is gathering the insight of researchers from around the world with respect to the future of space elevators. As scientific papers, reports and books are published, the research committee is pulling together this relative progress to assist academia and industry to progress towards an operational space elevator infrastructure.

ISEC is a traditional not-for-profit 501 (c) (3) organization with a board of directors and four officers: President, Vice President, Treasurer, and Secretary. inbox@isec.org / www.isec.org

Appendix C: Modern Day Space Elevator Body of Knowledge (www.isec.org)

Recently, a visitor to our International Space Elevator Consortium (ISEC) conference was quoted as saying, "You have a remarkable body of knowledge at www.isec.org. He was referring to the efforts of many scientists, engineers, and project/program professionals over the last 10 to 12 years. The leap in quality and currency shows that the Space Elevator is indeed twenty years beyond Dr. Edwards' breakthrough accomplishment saying, "it can be done." What is amazing are the conclusions from this body of knowledge:

1. Space Elevators are ready to initiate a developmental program
2. The tether material has been produced in the laboratory for the needed strength (150 GPa) and continuous length (1 meter per minute production) (note; not both capabilities at once - yet). This 2D material will be ready for the development team.
3. Space Elevators enable Missions off-planet with robust cargo movement as a complementary access to space with rockets.
4. Space Elevators are environmentally friendly in operations and enable Space Based Solar Power to eliminate hundreds of coals burning plants.
5. Space Elevators are prepared to be a partner in raising logistics to GEO and beyond. This concept is called the Dual Space Access Strategy with Advanced Rockets carrying people through the radiation belts while supporting LEO and MEO missions while Space Elevator do the heavy lifting to GEO and beyond with a very efficient and low-cost permanent transportation infrastructure.

ISEC is particularly proud of its most recent 18 month-long studies such as "Space Elevators are the Transportation Story of the 21st Century," "Space Elevators: The Green Road to Space, and "Leverage Dual Space Access Architecture – Advanced Rockets and Space Elevators." These study reports places Space Elevators into the near future and show how they support critical missions. One such mission is the enabling of Space Based Solar Power. This mission will lead to a much cleaner global environment by eliminating hundreds (or thousands) of coals burning plants. These reports also show how to support Mars colonies and Lunar villages by supplying their cargo. In addition, the Transportation report illustrates research accomplished by ISEC with Arizona State University showing the strengths of Space Elevators for interplanetary missions. Can you imagine 61 days to Mars? How about daily departures to Mars (no 26 month wait)? In addition, Space Elevators enable a tremendous benefit with massive cargo movement (170,000 tonnes per year to GEO and beyond). All this is accomplished with the Space Elevator architecture as a complement to rockets. This Dual Space Access Architecture (rockets and Space Elevators) is complementary and compatible - not competitive.

C.1: Body of Knowledge - Current

The principal source for the following information is at **www.isec.org**.

C.1.1   ISEC Studies

Latest engineering, management, operations, and developmental issues addressed in year-long studies conducted by Space Elevator experts. Download all 14 of these ISEC study reports in pdf for free **www.isec.org**.

    2023   Leverage Dual Space Access Architecture - Advanced Rockets and Space Elevators
    2021   The Climber-Tether Interface of the Space Elevator
    2021   Space Elevators: The Green Road to Space
    2020   Space Elevator is the Transportation Story of the 21st Century
    2020   Today's Space Elevator Assured Survivability Approach for Space Debris
    2019   Today's Space Elevator
    2018   Design Considerations for a Multi-Stage Space Elevator
    2017   Design Considerations for a Software Space Elevator Simulator
    2016   Design Considerations for Space Elevator Apex Anchor and GEO Node
    2015   Design Considerations for a Space Elevator Earth Port
    2014   Space Elevator Architectures and Roadmaps
    2013   Design Considerations for a Space Elevator Tether Climber
    2012   Space Elevator Concept of Operations
    2010   Space Elevator Survivability, Space Debris Mitigation
    And
    2017   Space Elevator: A History

C.1.2: International Academy of Astronautics Studies:

In addition, there were three other major studies conducted on the modern Space Elevator, by the International Academy of Astronautics and the Obayashi Corporation.
    2019   The Road to the Space Elevator Era - IAA International Academy of Astronautics (https://iaaspace.org) and
    2014   Space Elevators: An Assessment of the Technological Feasibility and the Way Forward - IAA International Academy of Astronautics (https://iaaspace.org)

C.1.3: Obayashi Corporation:

2014   The Space Elevator Construction Concept (https://www.obayashi.co.jp/en/news/detail/the_space_elevator_construction_concept.html)

C.1.4: References and Citations

References and Citations are listed by major topics (over 850 titles available).

C.1.5: Recent Articles and Publications

Recent articles and publications are shown in multiple locations with a special pair of publications that highlight the Modern-Day Space Elevator:

Journal of the British Interplanetary Society, Volume 76 No.7 July 2023
Special Issue: Future Directions for Space Elevators in PDF format.

- Raitt, David: Space Elevators: Introduction to This Special Issue.

- Wright, Dennis H.: Building the Space Elevator Tether.
- Bartoszek, Larry & Dennis H. Wright: Payload Design for the Space Elevator Climber.
- Knapman, John M.: Innovation and Research for Space Elevators.
- Swan, Peter A. & Cathy W. Swan: Huge Fast Spacecraft Travelling Our Solar System.
- Woods, Arthur, Andreas Vogler, Patrick Collins: The Lunar Space Elevator: A Key Technology for Realizing the Greater Earth Lunar Power Station.

SpaceFlight Vol 65 No 06 June 2023. In pdf format:

- **THE RIGHT STUFF:** The space elevator's tether could be built soon using lightweight, ultra-strong materials such as single crystal graphene, hexagonal boron nitride or carbon nanotubes. By Adrian Nixon, John Knapman and Dennis Wright. Diagrams courtesy ISEC/Adrian Nixon.
- **GALACTIC HARBOURS**: A galactic harbour is a permanent transportation infrastructure including two space elevators that will enable a space economy. By Michael Fitzgerald. Images courtesy Lux Virtual & Galactic Harbour Inc.
- **COOPERATION & COMPETITION**: A dual space access strategy leverages the best of both space elevators and rockets for a greener road to space. By Peter Swan.
- **MODERN DAY SPACE ELEVATORS**: The eighth space elevator architecture, as a permanent space transportation system, has remarkable transformative capabilities. By Cathy Swan.
- **INCREDIBLE ENGINEERING**: Japan's construction giant, Obayashi Corporation, has a vision for a space elevator that draws on NASA's earlier work. By Rob Coppinger.

C.1.6: ISEC Webinars

Recently, the modern Space Elevator has been discussed within webinars that are accessible on ISEC website as well as YouTube. They are:
- Oct 10, 2020 - Dual Space Access Architecture - Peter Swan (part of World Space Week)
- Aug 28, 2020 - Architectural Engineering for the Space Elevator - Michael Fitzgerald
- July 17, 2020 - How Space Elevators Work: Physics Concepts - Dennis Wright
- May 29, 2020 - Graphene: The Last Piece of the Space Elevator Puzzle? - Adrian Nixon
- Apr 30, 2020 - Today's Space Elevator - Peter Swan

C.1.7: YouTube Videos

Youtube has 49 videos from ISEC on space elevators at:
https://www.youtube.com/@internationalspaceelevator8857/videos

www.ingramcontent.com/pod-product-compliance
Lightning Source LLC
Chambersburg PA
CBHW080943170526
45158CB00008B/2355